# SpringerBriefs in Applied Sciences and Technology

## Nonlinear Circuits

W0036922

**Series editors**

Luigi Fortuna, DIEEI, University of Catania, Catania, Italy
Guanrong Chen, Heidelberg, Baden-Württemberg, Germany

*SpringerBriefs in Nonlinear Circuits* promotes and expedites the dissemination of substantive new research results, state-of-the-art subject reviews and tutorial overviews in nonlinear circuits theory, design, and implementation with particular emphasis on innovative applications and devices. The subject focus is on nonlinear technology and nonlinear electronics engineering. These concise summaries of 50–125 pages will include cutting-edge research, analytical methods, advanced modelling techniques and practical applications. Coverage will extend to all theoretical and applied aspects of the field, including nonlinear electronic circuit dynamics from modelling and design to their implementation. Topics include but are not limited to:

- nonlinear electronic circuits dynamics;
- oscillators;
- cellular nonlinear networks;
- arrays of nonlinear circuits;
- chaotic circuits;
- system bifurcation;
- chaos control;
- active use of chaos;
- nonlinear electronic devices;
- memristors;
- circuit for nonlinear signal processing;
- wave generation and shaping;
- nonlinear actuators;
- nonlinear sensors;
- power electronic circuits;
- nonlinear circuits in motion control;
- nonlinear active vibrations;
- educational experiences in nonlinear circuits;
- nonlinear materials for nonlinear circuits; and
- nonlinear electronic instrumentation.

**Contributions to the series** can be made by submitting a proposal to the responsible Springer contact, Oliver Jackson (oliver.jackson@springer.com) or one of the Academic Series Editors, Professor Luigi Fortuna (luigi.fortuna@dieei.unict.it) and Professor Guanrong Chen (eegchen@cityu.edu.hk).

**Publishing Ethics:** Researchers should conduct their research from research proposal to publication in line with best practices and codes of conduct of relevant professional bodies and/or national and international regulatory bodies. For more details on individual ethics matters please see: https://www.springer.com/gp/authors-editors/journal-author/journal-author-helpdesk/publishing-ethics/14214

More information about this series at http://www.springer.com/series/15574

Fadhil Rahma · Saif Muneam

# Memristive Nonlinear Electronic Circuits

## Dynamics, Synchronization and Applications

 Springer

Fadhil Rahma
Department of Electrical Engineering
University of Basrah
Basrah, Iraq

Saif Muneam
Department of Electrical Engineering
University of Basrah
Basrah, Iraq

ISSN 2191-530X                ISSN 2191-5318  (electronic)
SpringerBriefs in Applied Sciences and Technology
ISSN 2520-1433                ISSN 2520-1441  (electronic)
SpringerBriefs in Nonlinear Circuits
ISBN 978-3-030-11920-1        ISBN 978-3-030-11921-8  (eBook)
https://doi.org/10.1007/978-3-030-11921-8

Library of Congress Control Number: 2018968370

This Springer imprint is published by the registered company Springer Nature Switzerland AG
The registered company address is: Gewerbestrasse 11, 6330 Cham, Switzerland

# Preface

Nonlinear systems have been receiving more interest in the design and circuit implementation for generating complex dynamics. In consequence, the nonlinear circuits are widely used in many applications such as secure communications, medicine, control, and so on. A memristor is a nonlinear element which makes the electronic circuits nonlinear and produces complex dynamics. Note, inserting the memristor in the nonlinear electronic circuits leads to extend the dimension of circuits, and the higher-dimensional circuits are more suitable for secure communication.

A new memristor model is proposed using *inverse tangent* function. The proposed model achieved the characteristics of the memristor (three fingerprints). The new function can be implemented by a bipolar transistor differential pair. Memristor and time delay are potential candidates for constructing new circuits with complex dynamics. The new memristive time-delay system based on the proposed memristor model is designed for obtaining a time-delay memristive differential equation. The system generates $n$-scroll chaotic attractor by adjusting the proposed nonlinear function. Also, a new time-delay memristive system excited by staircase function (nonautonomous) is proposed. It can generate a new family of grid of scrolls ($n \times m$ scroll). In addition, a new five-dimensional (5D) autonomous system which includes two memristors is introduced. The results of proposed systems and corresponding designed electronic circuits are shown through the numerical simulations and OrCAD PSpice.

The synchronization is considered as the basic idea of the nonlinear electronic circuit applications. In this book, two synchronization methods are applied to the memristive circuits: The first is PC synchronization, and the second is feedback control. The results show that the performance of feedback control is better than PC, so the feedback control is used in the application of cryptography. The cryptography technique is used for encrypting and decrypting a message. Two types of messages (analog: sound and digital: PN code) have been introduced for secure communications applications. There are two encryption stages: The first is key-stream, and the other is the masking.

Nowadays, the development of memristive electronic circuits is far from its end. The aim of this book is to confirm the memristor characteristics for proposing a new memristive model. Then, realization of electronic circuits that related to the proposed model. Design the synchronization scheme between the nonlinear circuits and using it for the application in secure communications.

The book is organized as follows. Chapter 1 provides a general introduction and applications of the memristor. Chapter 2 presents the theory of memristor and introduces some fundamental circuitry properties of the memristor. A new memristor model using *inverse tangent* function is proposed and realized using PSpice software. Chapter 3 introduces a proposed time-delay memristive system based on memristor model. Also, it proposed a new nonautonomous time-delay memristive system. A new five-dimensional (5D) autonomous system with two memristors is constructed. Chapter 4 presents two synchronization methods and applied on the memristive time-delay chaotic circuits. Chapter 5 proposes the cryptography technique for the encryption and decryption purposes. Finally, concluding remarks are drawn in Chap. 6.

We would like to thank **Full Prof. Dr. Luigi Fortuna, Dr. Mattia Frasca and Dr. Arturo Buscarino** from University of Catania, Italy, for many helpful discussions during various stages of this work.

Basrah, Iraq                                                                              Fadhil Rahma
April 2018                                                                                Saif Muneam

# Contents

# Chapter 1
# Introduction

## 1.1 What Is a Memristor?

Any two-terminal element exhibits *pinched hysteresis loop* when derived by bipolar periodic current (response voltage) or periodic voltage (response current), in the *voltage–current* plane, can be called *memristor*. The *pinched hysteresis loop* is a fingerprint of memristor [1].

In 1801, Sir Humphry Davy was searched for developing an electrical lighting device that would replace the diffusing gas lamps, Davy was built two-terminal electrical device has two oppositely directed sharply honed carbon rods, separated by a narrow air gap. He connected 1000 batteries in series (barely a year after Alessandro Volta invented the battery) for applying high DC voltage across carbon rods, thereby heralding the first world's electric lamp had discovered. Until only very recently, Davy's arc lamp had been misidentified as a light-emitting resistor. At the University of Hong Kong, [2], were built Davy's carbon rod arc device and know it is in fact a memristor by observing the voltage input and current response are in phase, and the Lissajoux figure of the two signals (voltage–current) shows the "pinched hysteresis loop" [2].

Leon Chua was discovered the memristor at the University of California, Berkeley in 1971. He was using the operational amplifiers and transistors to implement the element. The dynamic range of all currents and voltages and the operating frequency dependent on the operational amplifiers and transistors. Although the memristor is a passive device, Chua's memristor is active because of the internal power supplies used in laboratory models [3].

Before the memristor's discovery, there are nine types of elements: four active and five passive, which is used to implement the electronic circuits. Each two-terminal element (capacitor discovered in 1745, the resistor 1827, the inductor 1831) was defined by a relation between the two-state variables of the circuit: voltage, current, flux, charge. Figure 1.1 shows the general classification of the elements (active and passive).

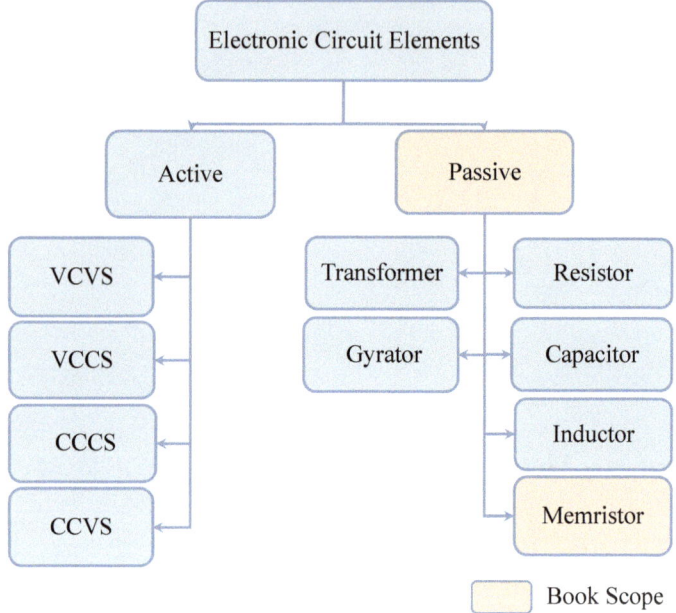

**Fig. 1.1** Taxonomy diagram of the circuit elements

## 1.2   The Three Circuit Elements

The electronic circuit is implemented by three fundamental two-terminal elements, which are known as:

1. Capacitor
2. Resistor
3. Inductor.

these two-terminal elements represent the relations between pairs of four fundamental circuit state variables: voltage $v$, current $i$, magnetic flux $\varphi$, and charge $q$ as shown in Fig. 1.2. The capacitor is the relation between voltage and charge $dq = Cdv$, the resistor is the relation between voltage, and current $dv = Rdi$ and the inductor is the relation between flux and current $d\varphi = Ldi$.

Leon Chua compared the four fundamental circuit variables (voltage, current, charge, flux) and the relationship between them with Aristotle's theory of matter [4]. According to Aristotle's theory, all matter consisted of the following four elements as shown in Fig. 1.3:

1. Earth
2. Water
3. Air
4. Fire.

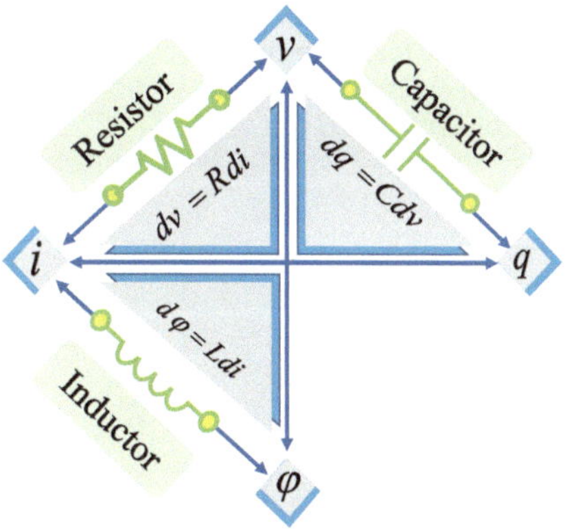

**Fig. 1.2** Relationship between the four-state circuit variables, before the invention of Leon Chua in 1971

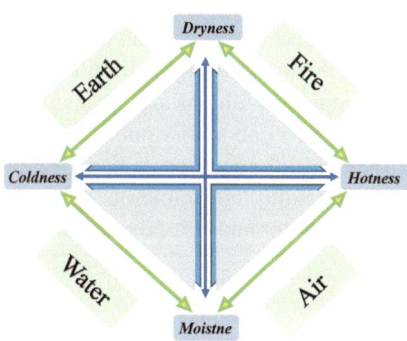

**Fig. 1.3** Aristotle's theory of matter

each one of these elements represents the relation between two of the four fundamental properties: dryness, hotness, moistness, and coldness. Chua was observed an amazing resemblance between the relation among elements and the relation among various circuit variables. To complete the symmetry, there should be a missing fourth fundamental circuit element, which is connected between flux and charge.

## 1.3 The Missing Element

Chua noticed the missing relation. Therefore, for the sake of completeness and symmetry, he predicted there should be a fourth fundamental circuit element. He called this hypothetical element, "memristor". Figure 1.4 shows the complete relation between all four circuit elements and variables. The memristor is *charge-controlled* or *flux-controlled* as illustrate in the equations below:

$$d\varphi = M(q)dq \tag{1.1}$$

$$dq = W(\varphi)d\varphi \tag{1.2}$$

**Fig. 1.4** Relationship
between the four-state circuit
variables, after the invention
of Leon Chua in 1971

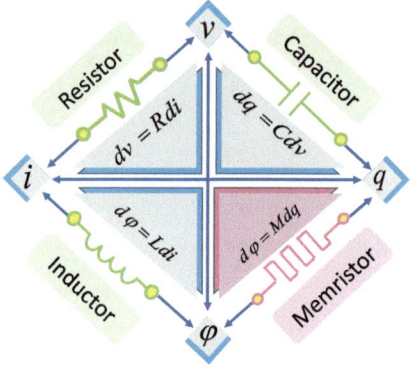

The above two equations will be explained in detail in Chap. 2.

The memristor roughly didn't show in the textbook or literature until May 2008, when Stanley Williams and his team at HP (Hewlett Packard) Information and Quantum Systems Laboratory published the milestone discovery, which is proved that the $TiO_2$ memristor device satisfied the pinched hysteresis loop. After Stanley Williams' group breakthrough, numerous papers have been published as shown in Fig. 1.5, these papers aim to analyze, fabricated and discussed the applications of the memristor [4–6].

## 1.4   The Memristor Applications

After HP discovered the passive model of the fourth fundamental circuit element (memristor), researchers from all the world have started significant experiments to

**Fig. 1.5** Publications for
memristor and memristive
system

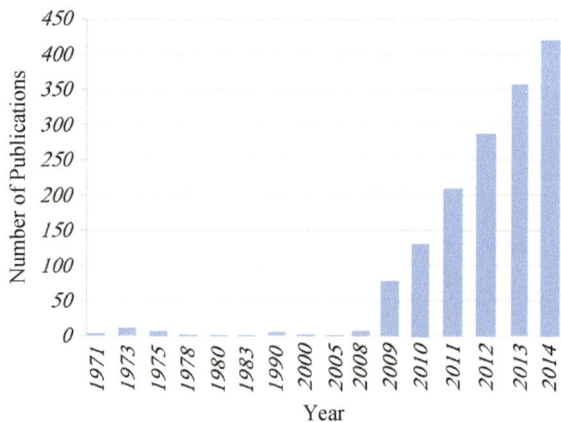

demonstrate the applications of the memristor. Memristors have been proposed in a wide range of applications such as chaotic electronic circuits, nonlinear analog circuits design, nonvolatile memory, and neuromorphic systems as shown in Fig. 1.6.

## 1.5  Nonlinear Circuits

Linear system theory provides an inadequate characterization of sustained oscillation in nature. So, the researchers moved from linear concepts and simple harmonics motion to nonlinear concepts. The nonlinear circuits can be realized by using resistors, capacitors, inductors, operational amplifiers and several circuits require analog multipliers. Each component is inexpensive except the analog multiplier. In 1963, Edward Lorenz invented the butterfly effect, the researchers tried to implement the electronic circuit of the Lorenz model. In 1983, Chua traveled to Japan via special research fellowship at Waseda University. While his arrival there, Chua was invited to observe the work of Prof. Matsumoto and his research group about a realization Lorenz system which was must be the first successful electronic circuit. The implemented circuit did not work because simple reason, at that time, the analog multiplier available had neither sufficient dynamic range nor ideal characteristics

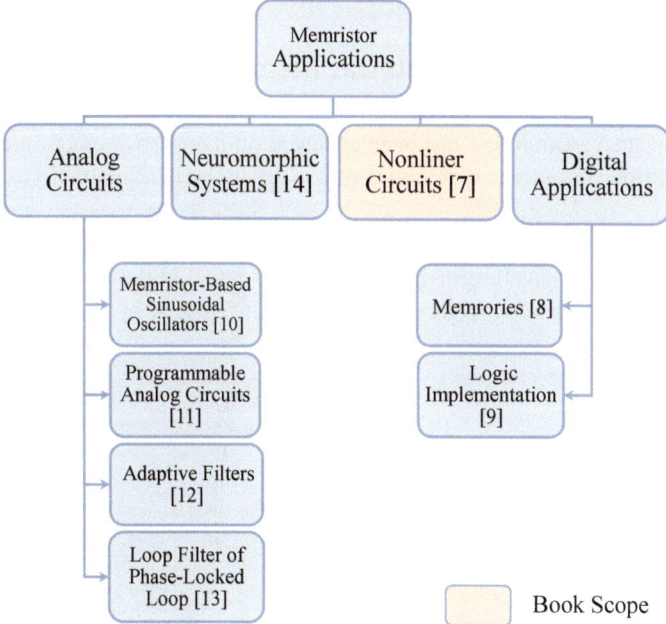

**Fig. 1.6**  Classification of memristor-based applications

required. All researchers believed that the special sets of nonlinear equations can show the chaos phenomena but only in computer simulation. So, Chua was hoped to dispel this idea [7].

Chua understood the mechanism of Lorenz circuit, and he could design simple nonlinear circuit consists of two capacitors ($C_1$, $C_2$), resistor ($R$), inductor ($L$) and nonlinear resistor (NR) as shown in Fig. 1.7. Existence of the nonlinear resistor (Chua's diodes) in Chua's circuit makes the circuit nonlinear. Chua's circuit is the simplest autonomous circuit can exhibit chaotic behavior. Before Chua's circuit invention, the electronics theory has been focusing on linear elements while after inventing it, the nonlinear circuits become in forefront [7].

The $v$–$i$ graph of memristor shows nonlinear characteristics (pinched hysteresis loop) as shown in Fig. 1.8. Because of the nonlinearity of the memristor, Chua suggested to replace Chua's diodes with memristors to produce nonlinear dynamics and chaos. Due to the random nature of chaotic circuits and nonlinearity of the memristor, the memristive chaotic circuits are well suitable to secure communication applications such as random number generation and encryption. In secure communications, the memristive chaotic circuits that have high frequency are immense potential for applications. One possibility for producing high-frequency chaotic circuits is by using memristor element. Inserting the memristor element in the electronic circuits will extend the dimensions of circuits. This increasing in dimensionality has implications for designing better secure communication systems [8–10].

## 1.6  A Brief Survey on Nonlinear Memristive Systems

A history of the research that has been achieved on memristive electronic circuits, synchronization, and secure communication is discussed in three tracks as follows:

**Track 1: Memristive Electronic Circuits**

- Itoh and Chua [9] have inserted the memristive model in the nonlinear oscillator system for the first time. Chua's diode had been replaced by the memristor, and

**Fig. 1.7** Chua's circuit realization

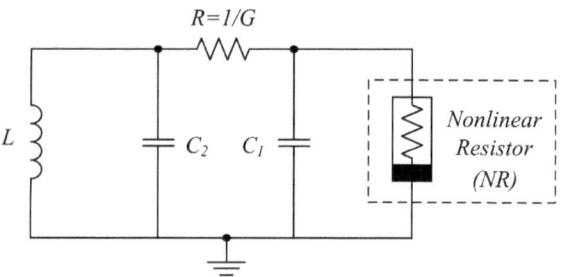

**Fig. 1.8** Current–voltage characteristic of the memristor (pinched hysteresis loop)

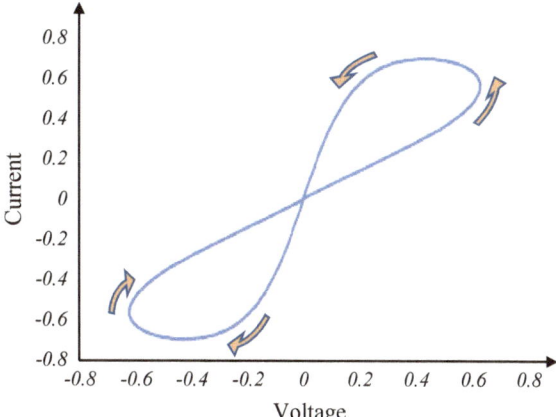

the memristor was characterized by the monotone-increasing and piecewise linear function. Then, concluded that the memristor are useful for designing nonlinear circuits.

- Muthuswamy and Kokate [10] have proposed memristor-based nonlinear circuits from canonical Chua's circuits, for the first time. They have been discussed three possible memristor-based electronic circuits and they discovered by simple "dimension extension" the memristor-based chaotic circuit can be obtained from the canonical Chua's circuit and by using only one negative element they could simplify the four element memristor-based chaotic circuit.
- Muthuswamy [11] provided a practical implementation of memristor-based chaotic circuits on breadboard, where the cubic nonlinearity has been used to perform the memristor characteristics. The memristive circuit has realized with off-the-shelf components such as capacitors, resistors, op-amp, and analog multipliers. Via using an analog integrator to obtain the electric flux across the memristor and then use the flux to obtain the memristor's characteristic function.
- Cheng et al. [12] proposed practical equivalent circuit of an active flux-controlled memristor. The constitutive relation of the memristor was smooth piecewise quadratic nonlinear. The dynamics of the implemented memristive circuit have been investigated via phase portraits, bifurcation diagrams, and Lyapunov exponents. Simulation and experimental results proved the proposed equivalent circuit realization of the active flux-controlled memristor.
- Li et al. [13] presented a practical implementation of a memristor-based chaotic system. The system has been generated by replacing the nonlinear resistor in the canonical Chua's circuit with a flux-controlled memristor characterized by a smooth continuous cubic monotone-increasing nonlinearity, in addition to nonlinearity, there is one negative element. The existence of nonlinear dynamics has been verified by computer simulations, Lyapunov exponents and bifurcation analysis.
- Pham et al. [14] have proposed two different autonomous nonlinear circuits based on memristive time-delay system (MTDS), (the six-element MTDS and the two-

element MTDS). The chaos behavior has been observed in very simple circuit configuration. It has been demonstrated more general conclusion, i.e., that chaos can appear in a system made by a memory element and a time-delayed memristive element.

- Wang et al. [15] have discussed memristor model by using hyperbolic sine nonlinear characteristic. Based on Chua's circuit, a fifth-order chaotic circuit designed with two flux-controlled memristor models. The proposed system proved through numerical simulation, Lyapunov spectrum, and bifurcation diagram. The new system has its unique dynamical behavior.
- Xu et al. [16] introduced a nonautonomous second-order memristive chaotic circuit, which is consisted of sinusoidal voltage source, resistor, capacitor, and memristor. The memristor implemented by off-the-shelf electronics components. The numerical simulations and theoretical analysis have been used for investigating the dynamical behaviors. The circuit realization is performed for verifying the numerical results.

**Track 2: The Synchronization**

- Carroll and Pecora [17] have proposed technique for synchronizing chaotic systems with different initial conditions for the first time. They described that the two chaotic systems linking with the same signal or signals, so the trajectories of one of the systems will converge to the same values of the other. After the synchronization has been done, the two systems will remain in step with each other. These ideas have been applied to two well-known systems (Lorenz and Rössler) as well as the real electronic circuit for synchronization.
- Liao and Tsai [18] have designed adaptive synchronization of master–slave chaotic systems. An adaptive observer-based slave system has been introduced to overcome the problem of synchronization in the presence of unknown parameters and some disturbances. To demonstrate the effectiveness of the proposed scheme, they have been selected two well-known circuits (Rössler-like and Chua) as illustrative examples.
- Huang et al. [19] have investigated the synchronization of coupled time-delay chaotic systems with the presence of mismatch parameters by using feedback control. Lyapunov function used to prove quasi-synchronization criteria. The theoretical analysis of that criteria shows there is a strong constraint on the control duration; the delay of systems must be smaller than control duration.
- Tamba et al. [20] introduced six-term chaotic system by using absolute nonlinearity and then implemented the electronic circuit without using any analog multiplier. An adaptive control has been applied to get the synchronization between two of the reported chaotic systems, which are identical structures with unknown parameters and by choosing the Lyapunov function for designing error dynamical system.

**Track 3: The Secure Communication**

- Cuomo et al. [21] have described the realization of the Lorenz chaotic circuit. The synchronization chaotic circuit concept has been used two possible approaches introduced in secure communication by using Lorenz circuit in both transmitter and receiver sides. In the transmitter, the chaotic masking is added to the message, then regenerated the masking in the receiver side and subtracted it from the received signal (first approach). The second approach utilizes coefficients modulation techniques.

- Yang et al. [22] have proposed a scheme in secure communication-based chaotic system. The proposed scheme used two chaotic signals; first chaotic signal is used for synchronization purpose between transmitter and receiver Chua's circuit, and the second signal is used for encrypting the message by using multi-shift cipher. It concluded that the transmitted signal is not used for encrypting the message.
- Fallahi et al. [23] have presented secure communication method by using extended Kalman filter. For increasing the security of the data transmission, a chaotic communication method based on multi-shift cipher algorithm has been introduced. The key cipher is selected as one of the chaotic states, the key estimated is used to recover the message. For showing the effectiveness of the proposal, presented Chen system and compared with Lorenz and Genesio–Tesi systems.
- Volos et al. [24] have proposed autonomous chaotic circuit that belongs to jerk family of systems. The proposed circuit has one nonlinear (hyperbolic sine) term. The proposed system has been used in secure communication for encrypting sound signal. The scheme of encryption based on a random number generator. The performance of the proposed chaotic jerk circuit at encryption technique is assessed from the results of NIST-800-22 criterion.

## 1.7   Book Contributions

The aims of this book are fulfilled through the following contributions:

1. Proposing a new memristive model by using *inverse tangent function*, the circuit realization of the proposed model is based on current feedback operational amplifier (CFOA).
2. Proposing a developed analytical model for the memristive time-delay system in order to generate complex dynamics by using $n$-segment *inverse tangent function*.
3. Proposing a new memristive time-delay system for generating multi-scroll attractor signal in two dimensions (grid).
4. Proposing a new memristive system based on two memristors.
5. Two synchronization methods of Pecora and Carrol (PC) and feedback control method are adopted in order to faithfully recover the original signal.
6. Verifying the proposed memristive systems by encrypting information (sound and PN code), where indirect synchronization technique is used in order to offer secure communication.

# References

1. S.P. Adhikari, M.P. Sah, H. Kim, L.O. Chua, Three fingerprints of memristor. IEEE Trans. Circ. Syst. I Regul. Pap. **60**(11), 3008–3021 (2013)
2. L. Chua, If it's pinched it's a memristor. Semicond. Sci. Technol. **29**(10), 104001 (2014)
3. L. Chua, Memristor—the missing circuit element. IEEE Trans. Circ. Theor. **18**(5), 507–519 (1971)
4. A.G. Radwan, M.E. Fouda, *On the Mathematical Modeling of Memristor, Memcapacitor, and Meminductor*, vol. 26 (Springer International Publishing, Switzerland, 2015). ISBN 978-3-319-17490-7
5. J.J. Yang, M.D. Pickett, X. Li, D.A. Ohlberg, D.R. Stewart, R.S. Williams, Memristive switching mechanism for metal/oxide/metal nanodevices. Nat. Nanotechnol. **3**(7), 429 (2008)
6. D.B. Strukov, G.S. Snider, D.R. Stewart, R.S. Williams, The missing memristor found. Nature **453**(7191), 80 (2008)
7. M.P. Kennedy, Three steps to chaos. II. A Chua's circuit primer. IEEE Trans. Circ. Syst. I Fundam. Theor. Appl. **40**(10), 657–674 (1993)
8. A. Thomas, Memristor-based neural networks. J. Phys. D Appl. Phys. **46**(9), 093001 (2013)
9. M. Itoh, L.O. Chua, Memristor oscillators. Int. J. Bifurcat. Chaos **18**(11), 3183–3206 (2008)
10. B. Muthuswamy, P.P. Kokate, Memristor-based chaotic circuits. IETE Techn. Rev. **26**(6), 417–429 (2009)
11. B. Muthuswamy, Implementing memristor based chaotic circuits. Int. J. Bifurcat. Chaos **20**(05), 1335–1350 (2010)
12. B. Bo-Cheng, X. Jian-Ping, Z. Guo-Hua, M. Zheng-Hua, Z. Ling, Chaotic memristive circuit: equivalent circuit realization and dynamical analysis. Chin. Phys. B **20**(12), 120502 (2011)
13. Y. Li, L. Zhao, W. Chi, S. Lu, X. Huang, Implementation of a new memristor based chaotic system, in *IEEE Fifth International Workshop on Chaos-Fractals Theories and Applications*, pp. 92–96 (2012)
14. V.-T. Pham, A. Buscarino, L. Fortuna, M. Frasca, Simple memristive time-delay chaotic systems. Int. J. Bifurcat. Chaos **23**(04), 1350073 (2013)
15. Z. Wang, F. Min, and E. Wang, A new hyperchaotic circuit with two memristors and its application in image encryption. AIP Adv. **6**(9), 095316 (2016)
16. Q. Xu, Q. Zhang, B. Bao, and Y. Hu, Non-autonomous second-order memristive chaotic circuit. IEEE Access **5**, 21039–21045 (2017)
17. T.L. Carroll, L.M. Pecora, Synchronization in chaotic Systems, Phys. Rev. Lett. **64**(8), 215–248 (1990)
18. T.-L. Liao, S.-H. Tsai, Adaptive synchronization of chaotic systems and its application to secure communications. Chaos, Solitons & Fractals **11**(9), 1387–1396 (2000)
19. T. Huang, C. Li, W. Yu, G. Chen, Synchronization of delayed chaotic systems with parameter mismatches by using intermittent linear state feedback. IOPsci. Nonlinearity **22**(3), 569 (2009)
20. V.K. Tamba, K. Rajagopal, V.-T. Pham, D.V. Hoang, Chaos in a system with an absolute nonlinearity and chaos synchronization. Adv. Math. Phys. **2018** (2018)
21. K.M. Cuomo, A.V. Oppenheim, S.H. Strogatz, Synchronization of Lorenz-based chaotic circuits with applications to communications. IEEE Trans. Circ. Syst. II Analog Digital Signal Process. **40**(10), 626–633 (1993)
22. T. Yang, C.W. Wu, L.O. Chua, Cryptography based on chaotic systems. IEEE Trans. Circ. Syst. I Fundam. Theor. Appl. **44**(5), 469–472 (1997)
23. K. Fallahi, R. Raoufi, H. Khoshbin, An application of Chen system for secure chaotic communication based on extended Kalman filter and multi-shift cipher algorithm. Commun. Nonlinear Sci. Numer. Simul. **13**(4), 763–781 (2008)
24. C. Volos, A. Akgul, V.-T. Pham, I. Stouboulos, I. Kyprianidis, A simple chaotic circuit with a hyperbolic sine function and its use in a sound encryption scheme. Nonlinear Dyn. **89**(2), 1047–1061 (2017)

# Chapter 2
# The Memristor: Theory and Realization

Memristor is a two-terminal passive element. Even though the behavior of the memristor was investigated for two centuries ago, the concept of the memristor as the fourth circuit element was proposed by Leon Chua in 1971 [1]. It is a contraction of *memory* and *resistor*, since it can remember its previous state. Also, it can be defined as it is a thin-film electrical circuit element that changes its resistance depending on the amount of charge which flows through it. Regardless of the device material and physical operating mechanism, any two-terminal nonvolatile memory devices based on resistance switching are memristors. The behavior of the memristor cannot be achieved by any circuit using only the other three basic electronic elements (capacitor, resistor, inductor), so that proves the memristor is truly fundamental [2, 3].

This chapter explained the theory of memristor and introduced some fundamental circuitry properties of the memristor. A new memristor model using *inverse tangent* function has been proposed. The proposed model achieved the characteristics of a memristor (pinched hysteresis loop), and the results can be shown through the numerical simulation and the electronic circuit design using PSpice software. The observing results explained very good agreement between the experimental and numerical simulation.

## 2.1 Analogy of Memristor

Consider a resistance of memristor to be a pipe through which water flows. The electric charge is analogous to the water flows, so the flow rate of the water through the pipe is like electrical current, the voltage applied to the memristor is similar to the pressure at the input of the pipe, the diameter of the pipe is analogous to the resistor's obstruction to the flow of charge: Increasing the diameter of the pipe represents the decreasing of resistance and vice versa. The resistor has a fixed pipe diameter, while the memristor has the varied pipe diameter with the amount and the direction of the water flow. If the water flows through this pipe is in one direction, the diameter of the

© The Author(s), under exclusive license to Springer Nature Switzerland AG 2019
F. Rahma and S. Muneam, *Memristive Nonlinear Electronic Circuits*,
SpringerBriefs in Nonlinear Circuits, https://doi.org/10.1007/978-3-030-11921-8_2

pipe expands (becoming less resistive). If the water flows in the opposite direction, the diameter of the pipe shrinks making it more resistive. If the water pressure is turned off, the diameter of the pipe will freeze until the water is turned back on [4, 5].

## 2.2  Memristor Characteristics

An element can be considered as a memristor, when agree to the three fingerprints or properties [6]:

1. The first significant signature of the memristor is a pinched hysteresis loop.
2. Hysteresis loop area increases as frequency decreases.
3. For high frequency, the memristor will behave as a linear device like a resistor.

## 2.3  The Proposed Memristor Model

The memristor is a circuit element presenting a relationship between two fundamental circuit variables, the charge $q$ and the flux $\varphi$. Figure 2.1 shows the symbol of a memristor [7].

The charge $q$ and the flux $\varphi$ represent the time integral of element's current $i(t)$ and voltage $v(t)$, respectively:

$$q(t) \triangleq \int_{-\infty}^{t} i(\tau)d\tau \qquad (2.1)$$

$$\varphi(t) \triangleq \int_{-\infty}^{t} v(\tau)d\tau \qquad (2.2)$$

The memristor has been classified into two kinds: *charge-controlled* and *flux-controlled memristor* based on its constitutive relations as follows:

$$\varphi = \varphi(q) \qquad (2.3)$$

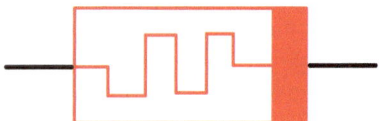

**Fig. 2.1**  Memristor symbol

$$q = q(\varphi) \tag{2.4}$$

where $\varphi(q)$ and $q(\varphi)$ are *continuous* and *nonlinear differentiable* functions [3].

Differentiating (2.3) and (2.4) with respect to time $t$, yields:

$$v = \frac{d\varphi}{dt} = \frac{d\varphi(q)}{dq} \frac{dq}{dt} = M(q)i \tag{2.5}$$

$$M(q) \triangleq \frac{d\varphi(q)}{dq} \tag{2.6}$$

where $M(q)$ is the memristance and has the Ohms ($\Omega$) unit.

$$i = \frac{dq}{dt} = \frac{dq(\varphi)}{d\varphi} \frac{d\varphi}{dt} = W(\varphi)v \tag{2.7}$$

$$W(\varphi) \triangleq \frac{dq(\varphi)}{d\varphi} \tag{2.8}$$

where $W(\varphi)$ is the memductance and has the Siemens (S) unit.

Memristance is a property of the memristor. The memristor built to express the characteristic of memristance. When the charge flows through the memristor in one direction, the memristor's resistance will increase, and if the flow of charge in the opposite direction in the memristor, the memristor's resistance will decrease. If the voltage applied to the memristor is turned off, the memristor will *remember* the last resistance that it had. When the flow of charge is starting again, the memristor's resistance will be last active [8].

Notice that (2.5) and (2.6) can be explained as *Ohm's law, except* that the resistance $M(q)$ at any time $t = t_0$ depends on the entire past history of $i(t)$ from $t = -\infty$ to $t = t_0$. Similarly, the memductance $W(\varphi)$ in (2.8) depends on the entire past history of $v(t)$ from $t = -\infty$ to $t = t_0$. It follows from (2.5) that the charge-controlled memristor defined in (2.3) is equivalent to the charge-dependent Ohm's law (2.5). Similarly, a flux-controlled memristor in (2.4) is equivalent to the flux-dependent Ohm's law (2.7). This book will consider the flux-controlled memristor.

Based on (2.4), a new proposed flux-controlled memristor model can be described by the following equation:

$$q(\varphi) = \frac{\varphi^2}{4} - \left( \varphi \tan^{-1}(\varphi) - \frac{1}{2} \ln\left(1 + \varphi^2\right) \right) \tag{2.9}$$

Hence, the conductance $W(\varphi)$ model of the flux-controlled memristor can be obtained by:

$$W(\varphi) = \frac{dq(\varphi)}{d\varphi} = \frac{\varphi}{2} - \tan^{-1}(\varphi) \tag{2.10}$$

The simulation results of $q(\varphi)$ and $W(\varphi)$ are shown in Fig. 2.2a, b, respectively.

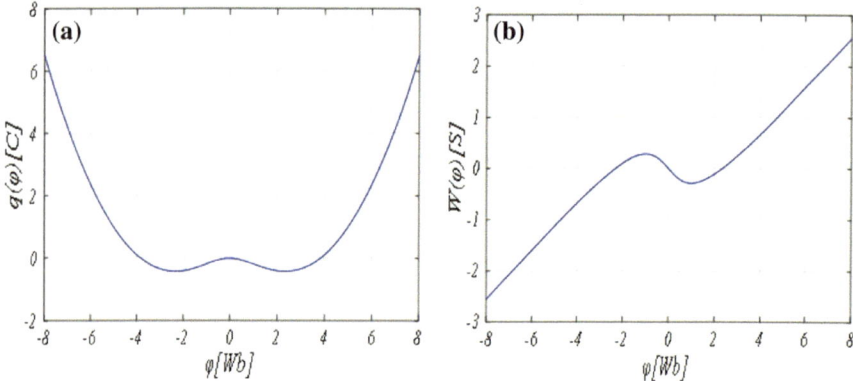

**Fig. 2.2 a** Memristor constitutive relation as a function of $\varphi$. **b** The relationship curve between conductance $W(\varphi)$ and flux $\varphi$

In 1976, Chua and Kang were extended the idea of the memristor, defined the memristor as a general nonlinear dynamical system called memristive system. Therefore, the memristor is a special case of a memristive system. According to that, the equations characterizing memristive system are described by [9]:

$$\begin{cases} \dot{x}_2 = F(x_2, x_1, t) \\ h = G(x_2, x_1, t)x_1 \end{cases} \tag{2.11}$$

where $x_1$, $h$, and $x_2$ denote the input, output, and state of the memristive system, respectively. The function $F$ is a continuously differentiable vector function and $G$ is a continuous scalar function.

According to the definition of the memristive system (2.11), the proposed memristive system using the trigonometric inverse tangent function (2.10) can be described as follows:

$$\begin{cases} \dot{x}_2 = x_1 - x_1 x_2 - ax_2 \\ h(x_2, x_1) = (bx_2 - \tan^{-1}(x_2))x_1 \end{cases} \tag{2.12}$$

where $a$ and $b$ are positive parameters.

To justify, whether the proposed model (2.12) is memristor or not, a periodic signal is applied as shown in Fig. 2.3. There are two cases:

**Case 1** Consider a sinusoidal voltage source as shown in Fig. 2.4a and defined by:

$$v_M = \begin{cases} A\sin(\omega t) & t \geq 0 \\ 0 & t < 0 \end{cases} \tag{2.13}$$

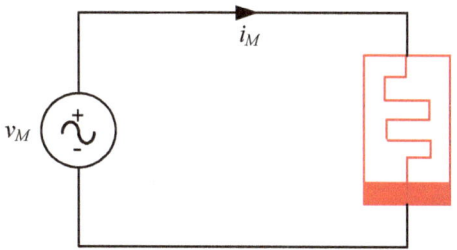

**Fig. 2.3** Memristive circuit

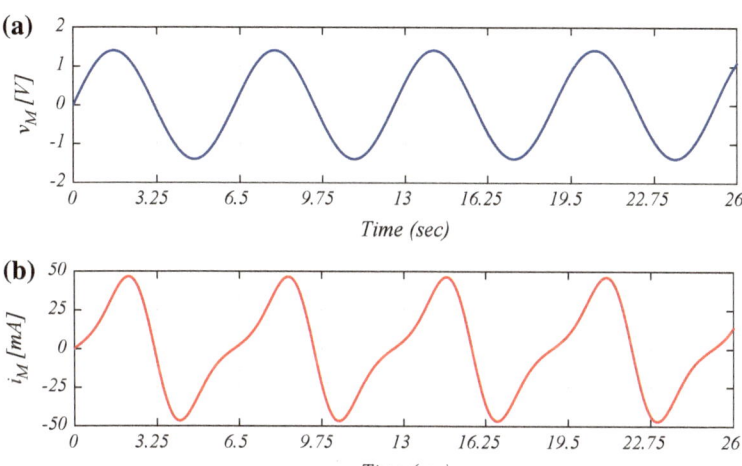

**Fig. 2.4** **a** Input voltage waveform, $A = 1.4$ V and $\omega = 1$ rad/s; **b** memristor current waveform

where $v_M$ is the memristor's voltage with and $\omega = 1$ rad/s. The memristor's current response $i_M$ can be determined from the constitutive relation (2.9). The corresponding flux can be determined from (2.2) as follows:

$$\varphi(t) = \int_0^t A\sin(\omega\tau)d\tau = \frac{A}{\omega}(1 - \cos(\omega t)), \quad t \geq 0 \tag{2.14}$$

Substituting (2.14) into (2.9), yields:

$$q(t) = -\left[\frac{A}{\omega}(1 - \cos(\omega t))\tan^{-1}\left(\frac{A}{\omega}(1 - \cos(\omega t))\right)\right.$$
$$\left. -\frac{1}{2}\ln\left(1 + \frac{A^2}{\omega^2}(1 - \cos(\omega t))^2\right)\right] + \frac{A^2}{4\omega^2}(1 - \cos(\omega t))^2 \tag{2.15}$$

Differentiating (2.15) with respect to $t$, yields:

$$i_M(t) = \left[ \frac{A}{2\omega}(1 - \cos(\omega t)) - \tan^{-1}\left( \frac{A}{\omega}(1 - \cos(\omega t)) \right) \right] A \sin(\omega t) \qquad (2.16)$$

The simulation result of $i_M$ is shown in Fig. 2.4b.

When the parameters of the proposed system (2.12) are set to: $a = 4$, $b = 0.5$ and $x_2(0) = 0.1$, the voltage–current relationship of the flux-controlled memristor, for sinusoidal input with different values of $\omega$, is depicted in Fig. 2.6. It proves that the pinched hysteresis loop characteristics of the memristor. When the frequency of the input signal is increased, the pinched hysteresis loop shrinks to a single-valued function as shown in Fig. 2.5 with $\omega = 20$ rad/s.

**Case 2** Consider the square voltage source as shown in Fig. 2.6a, and defined by:

$$v_M = \frac{4A}{\pi} \sum_{n=1}^{k} \frac{1}{n} \sin(0.3\pi n t) \qquad (2.17)$$

across the memristor with $A = 1.4$ V and $k = 1, 3, 5, \ldots$. The memristor current response can be determined from the constitutive relation (2.9) and given by:

$$q(t) = \frac{44.4A^2}{\pi^4} \sum_{n=1}^{k} \frac{(1 - \cos(0.3\pi n t))^2}{n^4}$$

$$- \left[ \frac{13.33A}{\pi^2} \sum_{n=1}^{k} \frac{(1 - \cos(0.3\pi n t))}{n^2} \tan^{-1}\left( \frac{13.33A}{\pi^2} \sum_{n=1}^{k} \frac{(1 - \cos(0.3\pi n t))}{n^2} \right) \right.$$

$$\left. -\frac{1}{2}\ln\left( 1 + \frac{178A^2}{\pi^4} \sum_{n=1}^{k} \frac{(1 - \cos(0.3\pi n t))^2}{n^4} \right) \right] \qquad (2.18)$$

**Fig. 2.5** Pinch hysteresis loop of the proposed memristive model (2.12) driven by sinusoidal stimulus. $A = 1.4$ V, $x_2(0) = 0.1$ and varying frequency

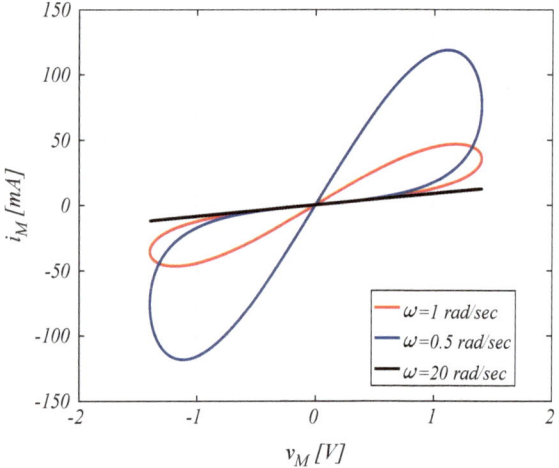

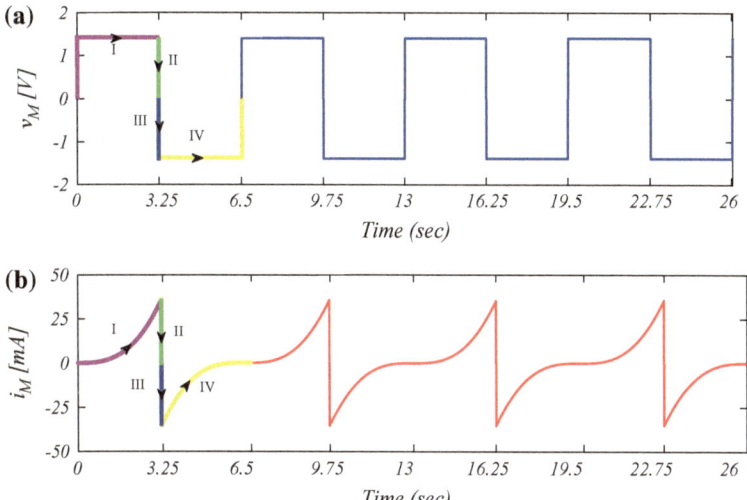

**Fig. 2.6** Memristor square wave excitation, response, and characteristic regions: **a** input voltage $A = 1.4$ V, $k = 99$, **b** memristor current

Differentiating (2.20) with respect to $t$, yields:

$$i_M(t) = \left[ \frac{6.67A}{\pi^2} \sum_{n=1}^{k} \frac{(1 - \cos(0.3\pi nt))}{n^2} \right.$$
$$\left. - \tan^{-1}\left( \frac{13.33A}{\pi^2} \sum_{n=1}^{k} \frac{(1 - \cos(0.3\pi nt))}{n^2} \right) \right] \frac{4A}{\pi} \sum_{n=1}^{k} \frac{\sin(0.3\pi nt)}{n} \qquad (2.19)$$

where $i_M$ is the current of the memristor as shown in Fig. 2.6b.

The system parameters are set to: $a = 4$, $b = 0.5$, $k = 99$ and initial condition $x_2(0) = 0.1$. The voltage–current relationship of the flux-controlled memristor, for square wave input, is plotted in Fig. 2.7. It's evidence that the pinched hysteresis loop characteristics of the memristor.

When the memristor input voltage $v_M$ is a square wave, the curves shown in Figs. 2.6 and 2.7 obviously demonstrate the memory effect. It is seen from the graph of both $v_M$ and $i_M$ in Fig. 2.6a and b. In the interval when $v_M$ is constant part I and part IV, the flux $\varphi$ follows a linearly decreasing or increasing course, which results in a similar effect on the memductance $W(\varphi)$ and consequently also on the current $i_M$. The memristor does not lose the value of $\varphi$ and $q$ when both $v_M$ and $i_M$ become zero at the instant when the power turns off. The current $i_M$ then abruptly change and the range of this change depends on the final $W(\varphi)$ value *remembered* by memristor.

**Fig. 2.7** Pinched hysteresis loop of the proposed memristive model (2.12) driven by square wave, $A = 1.4\,\text{V}$, $x_2(0) = 0.1$ and $k = 99$

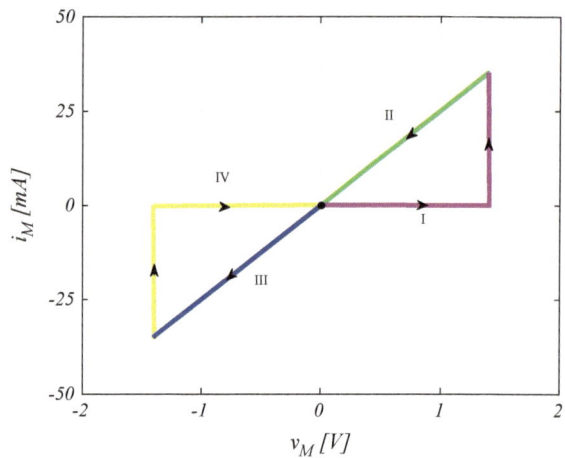

## 2.4   Circuit Realization of the Proposed Memristor Model

The memristor circuit design and implementation of the mathematical model (2.12) have been discussed in this part. The circuit design was achieved based on current feedback operational amplifier (CFOA). The CFOA has significant advantages over the conventional op-amp. These advantages include: (i) wider and nearly constant bandwidth at low/medium gains, (ii) relatively much higher slew rates (typically 2000 V/μs) while the conventional op-amp about (0.5 V/μs) for the most popular μA741, and (iii) less requirements of external passive components to perform a specified function. The CFOA has high bandwidth (around 60 MHz at gain of $-1$ and around 33 MHz at gain of $-10$) and provides excellent DC performance with very fast large signal response. It essentially as 4-terminal element (as opposed to the op-amp, which is 3-terminal element). Utilizing the CFOA is indeed much more versatile than op-amp in realizing a variety of configurations for a class of analog signal processing and signal generation circuits [10, 11].

The memristive system (2.12) is emulated by common off-the-shelf discrete components such as resistors, a capacitor, multipliers (AD633), current feedback operational amplifiers (AD844) and transistors (Q 2N3904). The proposed memristor circuit implementation shown in Fig. 2.8 includes two multipliers and five parts that are Part I: *Inverter* ($N_0$); Part II: *Summer* ($N_1$); Part III: *Integrator* ($N_2$); Part IV: *Inverse tangent function* ($N_3$; Part V: *Subtractor* ($N_4$). Note that the inverter $N_0$ multiplies the memristor input $X_1$ by negative gain. The part $N_1$ is summing the three terms (they have all negative gains), needed for the $X_2$ variable. The integration time constant of the integrator $N_2$ is determined by $R_7C$. The inverse tangent function block ($N_3$) is implemented using the suggested circuit as shown in Fig. 2.9. The circuit includes three parts; Part I: *Current mirror* ($P_1$); Part II: *Differential pair* ($P_2$);

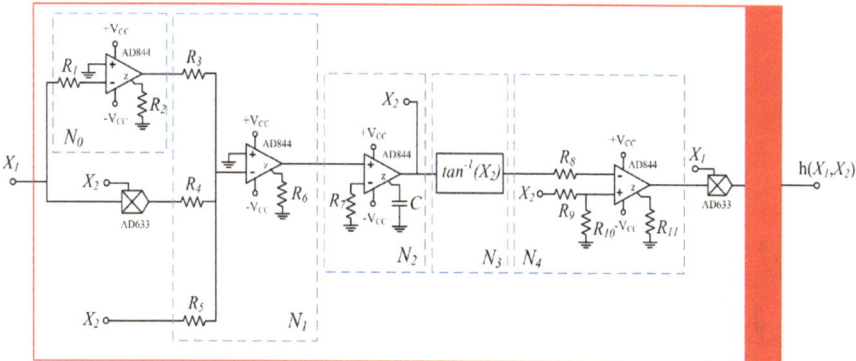

**Fig. 2.8** Circuital implementation which emulates the memristor device. The values of the components are selected as: $R_1 = R_2 = R_8 = R_{10} = R_{11} = 1\,\text{k}$, $R_9 = 3\,\text{k}$, $R_3 = R_6 = 2\,\text{k}$, $R_4 = 0.2\,\text{k}$, $R_5 = 0.5\,\text{k}$, $R_7 = 10\,\text{k}$, $C = 100\,\text{nF}$, $V_{cc}(\pm 12\,\text{V})$

**Fig. 2.9** Scheme of the circuit realization of the inverse tangent function. The values of the components are selected as:
$R_{P_1} = 1.2\,\text{k}$, $R_{P_2} = R_{P_3} = 1\,\text{k}$, $R_{P'_3} = 3\,\text{k}$, $V = -10.3\,\text{V}$, $V_{cc}(\pm 12\,\text{V})$

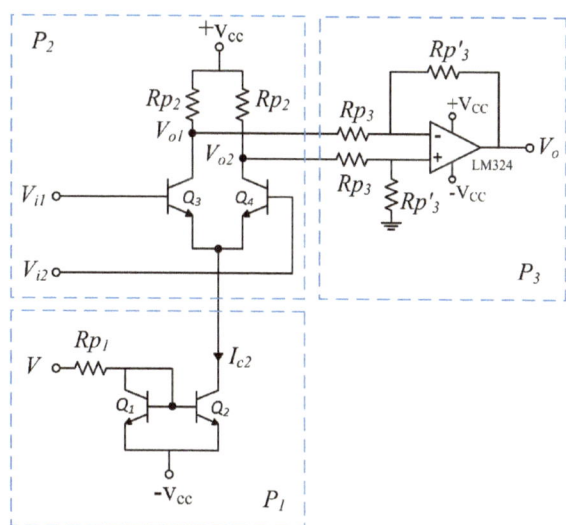

Part III: *Differential to single-ended converter* ($P_3$). In part I, ($P_1$), $V$ is the control voltage input which sets the current in the current mirror so that:

$$I_{c2} = \frac{(V - V_{\text{be1}} + |-V_{\text{CC}}|)}{Rp_1} \qquad (2.20)$$

where $I_{c2}$ is the collector current of the transistor $Q_2$.

The differential pair ($P_2$) amplifier is formed from two identical transistors $Q_3$ and $Q_4$ which are coupled at their emitters (emitter-coupled pair). When the input voltage $V_{i1}$ and $V_{i2}$ is applied, the differential output voltage is observed as follows:

$$V_{od} = V_{o1} - V_{o2} = (I_{c3} - I_{c4})R_{P_2} \tag{2.21}$$

$$I_{c2} = (I_{c3} + I_{c4}) \tag{2.22}$$

$$V_{od} = I_{c2}R_{P_2}\tan^{-1}\left(\frac{V_{i1} - V_{i2}}{2V_T}\right) \tag{2.23}$$

where $V_{od}$ is the differential output voltage, $I_{c3}$ and $I_{c4}$ are the collector currents of transistors $Q_3$ and $Q_4$, respectively, $V_T \cong 25\,\text{mV}$ at room temperature. When the result of $(V_{i1} - V_{i2})$ is greater than $3V_T$, the collector currents of transistors $Q_3$ and $Q_4$ are almost independent of the $(V_{i1} - V_{i2})$ because one of the transistors turns off and the other conducts. If the magnitude of $(V_{i1} - V_{i2})$ less than $V_T$, the circuit operated in approximately linear region. This circuit is differential input–differential output, so Part III $(P_3)$ has been used to convert the differential output to single-ended.

The input, output, and the internal state of the memristor are $X_1$, $h(X_1, X_2)$, and $X_2$, respectively. The voltage across the capacitor $C$ corresponding to the variable $X_2$ which is internal state of the memristive system. By applying Kirchhoff's circuit laws, the equation, of the circuit in Fig. 2.8 are derived as:

$$\begin{cases} \frac{dX_2(t)}{dt} = \frac{1}{R_7C}\left(\frac{R_2}{R_1}\frac{R_6}{R_3}X_1 - 0.1\frac{R_6}{R_4}X_1X_2 - \frac{R_6}{R_5}X_2\right) \\ h(X_1, X_2) = \left(\frac{R_{10}}{R_9+R_{10}}\left(1 + \frac{R_{11}}{R_8}\right)X_2 - \frac{R_{11}}{R_8}I_{c2}R_{P_2}\tan^{-1}\left(\frac{X_2}{2V_T}\right)\right)\frac{X_1}{10} \end{cases} \tag{2.24}$$

Equation (2.24) match Eq. (2.12) with $a = \frac{R_6}{R_5}$, $b = \frac{R_{10}}{R_9+R_{10}}\left(1 + \frac{R_{11}}{R_8}\right)$ and the inverse tangent function part match Eq. (2.23) with $V_{i1} = X_2$ and $V_{i2} = 0$ (connected to ground). The circuit depicted in Fig. 2.8 has been successfully implemented using the PSpice software, and thereby, the simulation results are shown in Figs. 2.10, 2.11, 2.12, and 2.13. The conductance shown in Fig. 2.10b has some negative parts of the curve; it implies the proposed memristor model is locally active, meaning that it can never be built without an internal power supply. Figures 2.11a, b and 2.12a, b show the time series of sinusoidal and square input voltages and the response currents. The voltages and currents have the same phases, which indicate that the proposed model is purely memristive without any series reactive element. The pinched hysteresis loop shown in Fig. 2.13a, b is proving that the proposed model is memristive system. The experimental results matched to the simulation results.

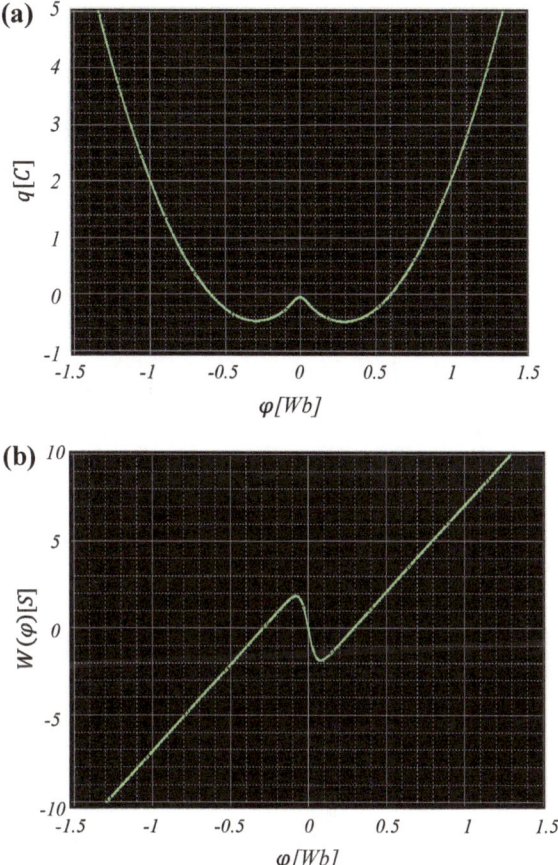

**Fig. 2.10** PSpice results, **a** the curve between $\varphi$ and $q$. **b** the relationship curve between flux $\varphi$ and memory conductance $W(\varphi)$

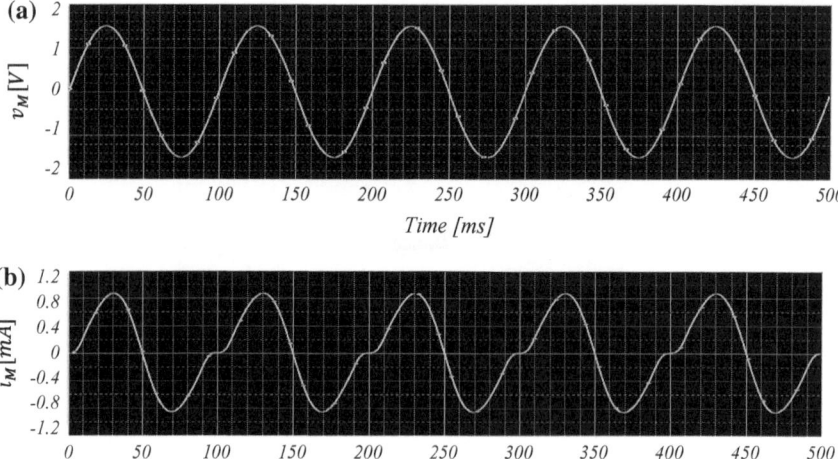

**Fig. 2.11** PSpice results, **a** time series sinusoidal input voltage $v_M$ applied to the memristor and **b** the response current $i_M$ through memristor

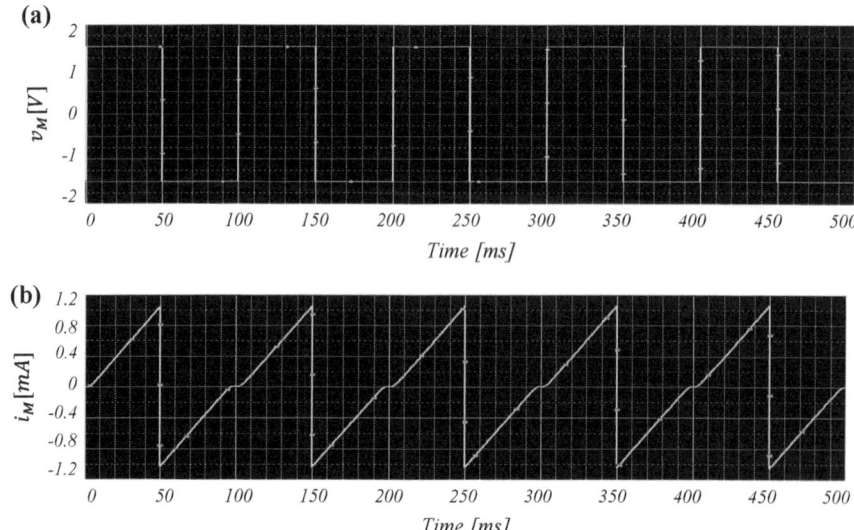

**Fig. 2.12** PSpice results, **a** time series square wave input voltage $v_M$ applied to the memristor and **b** the response current $i_M$ through memristor

**Fig. 2.13** PSpice results, pinch hysteresis loop of memristor model for; **a** sinusoidal wave input and **b** square wave input

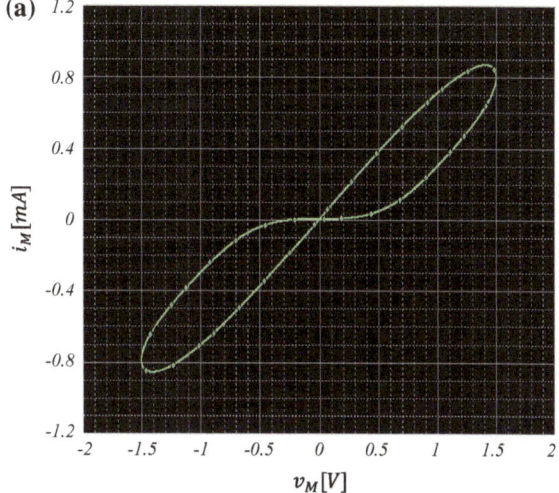

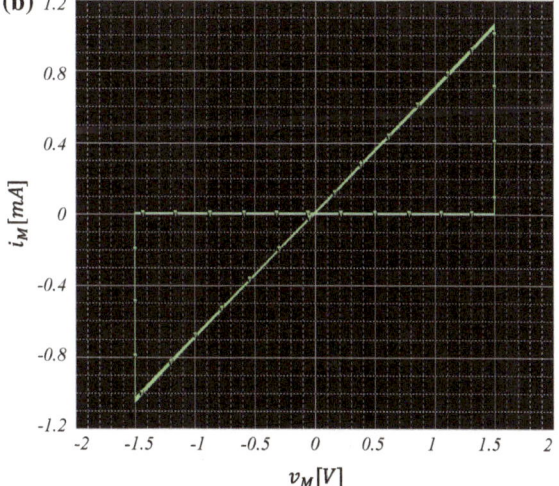

# References

1. T. Prodromakis, C. Toumazou, L. Chua, Two centuries of memristors. Nat. Mater. **11**, 478–481 (2012)
2. A. Buscarino, L. Fortuna, M. Frasca, and L. Valentina Gambuzza, A chaotic circuit based on Hewlett-Packard memristor. Chaos: Interdisc. J. Nonlinear Sci. **22**(2), 023136 (2012)
3. L. Chua, Resistance switching memories are memristors. Appl. Phys. Mater. Sci. Process. **102**, 765–783 (2011)
4. Z. Mustaqueem, *Analysis, design and modelling of memristor emulator and its application in FM detector circuit*. M.Sc. thesis, Jamia Millia Islamia New Delhi, 2016
5. R.S. Williams, How we found the missing memristor. IEEE Spectr. **45**, 3–16 (2008)
6. S.P. Adhikari, M.P. Sah, H. Kim, L.O. Chua, Three fingerprints of memristor. IEEE Trans. Circ. Syst. I Regul. Pap. **60**(11), 3008–3021 (2013)
7. B. Muthuswamy, L.O. Chua, Simplest chaotic circuit. Int. J. Bifurcat. Chaos **20**(05), 1567–1580 (2010)
8. L. Chua, Memristor-the missing circuit element. IEEE Trans. Circ. Theor. **18**(5), 507–519 (1971)
9. L. Chua, S.M. Kang, Memristive devices and systems. Proc. IEEE **64**(2), 209–223 (1976)
10. R. Senani, D. Bhaskar, A. Singh, V. Singh, *Current Feedback Operational Amplifiers and Their Applications* (Springer Science & Business Media, New York, 2013). ISBN 978-1-4614-5187-7
11. R. Senani, Realization of a class of analog signal processing/signal generation circuits: novel configurations using current feedback op-amps. Frequenz **52**(9–10), 196–206 (1998)

# Chapter 3
# Memristive Electronic Circuits

The researchers are looking for increasing the complexity of behavior while keeping the systems as simple as possible. The chaotic systems that have multi-scrolls attractors give more complex behaviors compared to chaotic systems of double scrolls. Generation of multi-scroll chaotic attractor is described in dynamical systems, and this can be shown in several applications such as encoding the fingerprint image, secure communication, dominating motion trends of autonomous mobile robots, encoding medical image, and procreating pseudo-random number (PRN). As a consequence, more attractive researches trend to generate a chaotic system with multi-scrolls which have appeared recently in both directions of theoretical and practical applications. Several researches were studied intensively the three-dimensional chaotic systems to generate multi-scrolls [1].

This chapter is organized as follows: First, it has been proposed a time-delay memristive system based on the nonlinear function (inverse tangent function) which generates $n$-scroll. Also, it proposed time-delay memristive system excited by staircase function (nonautonomous) which creates new families of scroll (grid chaotic attractor) to be generated. Second, a new five-dimensional (5D) autonomous system with two memristors is introduced. The dynamic mechanism is investigated by analyzing the system stability and bifurcation diagram. The results of the proposed systems can be shown through the numerical simulations, and the design of electronic circuits has been done using PSpice software.

## 3.1 Time-Delay Memristive System

Time-delay systems are described by delay differential equations (DDEs). Similar to ordinary differential equations (ODEs), the nature of solutions of DDEs can be obtained by performing a linear stability analysis of the equilibrium solutions. The major difference between ODEs and DDEs in the case of DDEs is that the phase space is infinite-dimensional, while in the state of ODEs, it is finite dimensional. Chaos

© The Author(s), under exclusive license to Springer Nature Switzerland AG 2019
F. Rahma and S. Muneam, *Memristive Nonlinear Electronic Circuits*,
SpringerBriefs in Nonlinear Circuits, https://doi.org/10.1007/978-3-030-11921-8_3

**Fig. 3.1** Structure of the
time-delay system

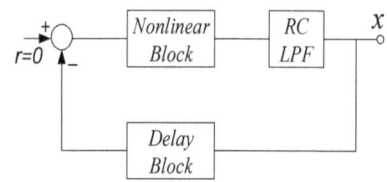

is obtained from any systems qualified by (one or more) nonlinear DDEs. Several
properties of DDEs made the analysis more difficult. The DDE can be defined as [2, 3]

$$\dot{x} = f(x(t), x(t - \tau)) \tag{3.1}$$

Because of the presence of the delay ($\tau$), the above system not only depends on
the current state of the system, but also on the state of the system at previous times.

The time-delay chaotic system is built from a simple feedback scheme with mini-
mum number of blocks required to observe a chaotic behavior in a dynamical circuit
with single state variable. It consists of three blocks: a nonlinearity, a $RC$ low-pass
filter (LPF), and a time-delay block. The scheme of the system is shown in Fig. 3.1 [4].

### 3.1.1  A New Autonomous Memristive Time-Delay System

Based on the memristor model (2.12) which is proposed in the previous chapter, the
mathematical model of the proposed new time-delay system is given as follows:

$$\begin{cases} \dot{x}_1 = F(x_{1\tau}) - a_1 h(x_1, x_2) \\ \dot{x}_2 = x_1 - x_1 x_2 - a_2 x_2 \end{cases} \tag{3.2}$$

where $a_1$ and $a_2$ are positive parameters, $\tau$ is the time delay, $x_\tau$ denotes $x(t-\tau)$, $F(x_{1\tau})$
is a nonlinear function, and $h(x_1, x_2)$ is the output of the memristor as presented in
(2.12).

- **The Proposed Nonlinear Function**

To generate $n$-scroll chaotic attractor from the system (3.2), the inverse tangent function is proposed, which describes the nonlinearity in the system (3.2). The nonlinear
function has two cases to generate chaotic attractor:

To create the chaotic attractor with an even number of scrolls, the adjustable
nonlinear function $F(x)$ is constructed as:

$$F(x) = F_1(x)$$

$$= -a_3 x + a_4 \sum_{k=-(M-1)}^{M-1} \tan^{-1}(gx + kr) \tag{3.3}$$

Similarly, to generate the chaotic attractor with an odd number of scrolls, the adjustable nonlinear function is constructed as:

$$F(x) = F_2(x)$$

$$= -a_3 x + a_4 \sum_{k=1}^{M} [\tan^{-1}(gx + kr) + \tan^{-1}(gx - kr)] \qquad (3.4)$$

where $a_3$, $a_4$, $g$, $r$, and $M$ are positive parameters. The nonlinear functions in (3.3) and (3.4) can offer $2M (M = 1, 2, \ldots)$ and $2M + 1$ scrolls in attractor, respectively.

## • Equilibria and Stability

The equilibrium points of the system (3.2) can be found by solving the following two equations:

$$\begin{cases} F(x_{1\tau}) - a_1 h(x_1, x_2) = 0 \\ x_1 - x_1 x_2 - a_2 x_2 = 0 \end{cases} \qquad (3.5)$$

Due to symmetry of the functions $F_1(x)$ and $F_2(x)$, one only needs to consider the case $x \geq 0$. Obviously, the system (3.2) with (3.3) and (3.4) and fixing the parameters as $a_1 = 0.1$, $a_2 = 4$, $a_3 = 1.82$, $a_4 = 0.6$, $\tau = 1$ can be classified into several regions based on the characteristic regions of nonlinear functions. Figure 3.2a, b, and c shows the regions of characteristic functions of 2-, 3-, and 4-scroll chaotic attractors, respectively, where $\pm D_i (i = 0, 1, 2, \ldots, M)$ denotes different regions and $\pm x_{ei} (i = 0, 1, 2, \ldots, M)$ are the equilibrium points. The scrolls are located around the equilibrium points. For the two-scroll chaotic attractor, there are three equilibrium points $x_{e0}$, $+x_{e1}$, $-x_{e1}$ as shown in Fig. 3.2a; $x_{e0}$ is unstable for all $\tau$, and the stability of $\pm x_{e1}$ is dependent on the value of $\tau$ [4]. For the three and four scrolls, there are five $x_{e0}$, $+x_{e1}$, $-x_{e1}$, $+x_{e2}$, $-x_{e2}$ and seven $x_{e0}$, $+x_{e1}$, $-x_{e1}$, $+x_{e2}$, $-x_{e2}$, $+x_{e3}$, $-x_{e3}$ equilibrium points as shown in Fig. 3.2b and c, respectively. The saddle points are unstable and independent on the value of $\tau$ such as $\pm x_{e1}$ shown in Fig. 3.2b and $x_{e0}$, $\pm x_{e2}$ shown in Fig. 3.2c. The remaining equilibrium points are unstable saddle-focus, which generate the scrolls.

Starting from the parameter values that introduced above and forcing on the two-scroll circuit, we now show some bifurcation diagram to highlight the role of two important parameters. Bifurcation diagrams, obtained plotting the local maxima of the state variable $x_1$, are shown in Figs. 3.3 and 3.4. The bifurcation diagram with respect to the time delay $\tau$ is depicted in Fig. 3.3. For $\tau < 0.83$ s, the equilibrium points are stable, and depending on initial conditions, the trajectory of the system (3.2) with nonlinearity (3.3) and $M = 1$, $a_1 = 0.1$, $a_2 = 4$, $a_3 = 1.82$, $a_4 = 0.6$, and $g = 10$ converges, and no limit cycle originates in this situation. Increasing $\tau$, $\tau \geq 0.83$ s, a limit cycle is predicted and starts to interact with unstable equilibrium points, and so chaos emerges.

**Fig. 3.2** $F(x)$ and its characteristics regions of chaotic $n$-scroll; **a** two-scroll attractor, **b** three-scroll attractor, and **c** four-scroll attractor

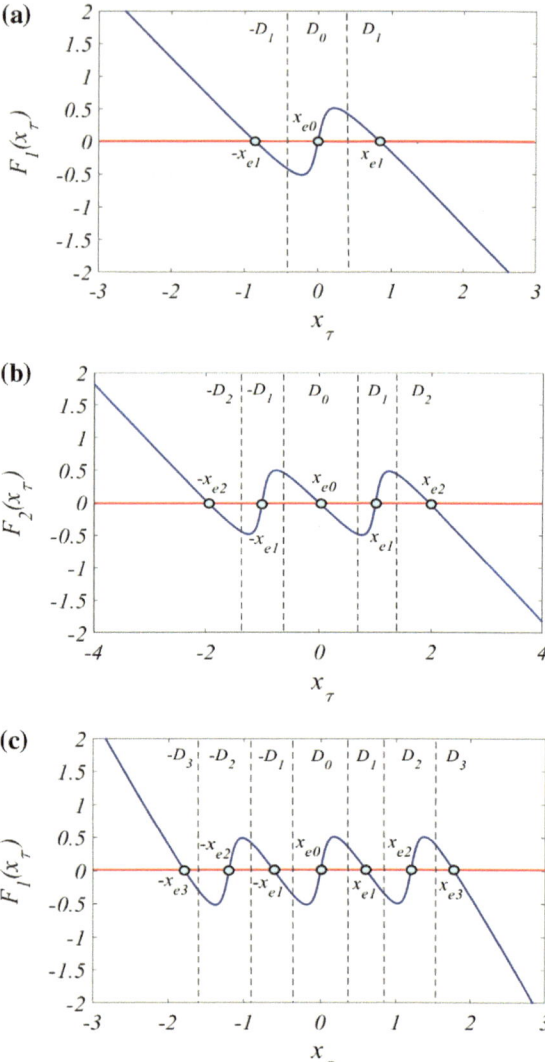

The bifurcation diagram with respect to parameter $a_3$ is depicted in Fig. 3.4, the delay is fixed at $\tau = 1$ s, and the bifurcation parameter $a_3$ is varied to explore the system dynamics. For $a_3 \geq 1.65$, the fixed point loses its stability through Hopf bifurcation, and a stable limit cycle emerges. When $a_3 = 1.67$, the limit cycle period 1 becomes unstable and a period 2 cycle appears. Further period doubling occurs at $a_3 = 1.71$ (period 2 to period 4). Through period doubling bifurcation, the system enters the chaotic behavior when $a_3 = 1.76$.

**Fig. 3.3** Bifurcation diagram for the memristive time-delay system (3.2) with (3.3) when $a_1 = 0.1$, $a_2 = 4$, $a_3 = 1.82$, $a_4 = 0.6$, $M = 1$, $r = 4$, and $g = 10$, the initial conditions, $(x_1(0), x_2(0)) = (0.1, 0.1)$, and $\tau$ as a varying parameter

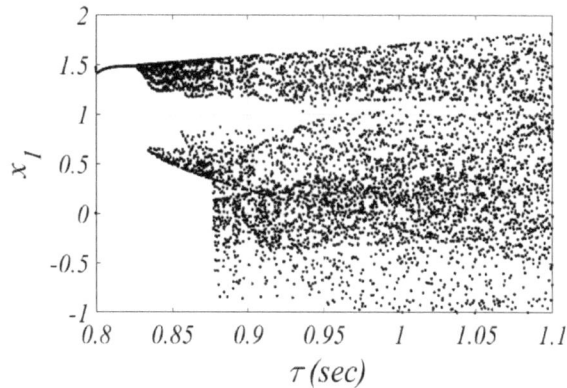

**Fig. 3.4** Bifurcation diagram for the memristive time-delay system (3.2) with (3.3) when $a_1 = 0.1$, $a_2 = 4$, $a_4 = 0.6$, $g = 10$, $\tau = 1$, $r = 4$ and $M = 1$, the initial conditions, $(x_1(0), x_2(0)) = (0.1, 0.1)$, and $a_3$ as a varying parameter

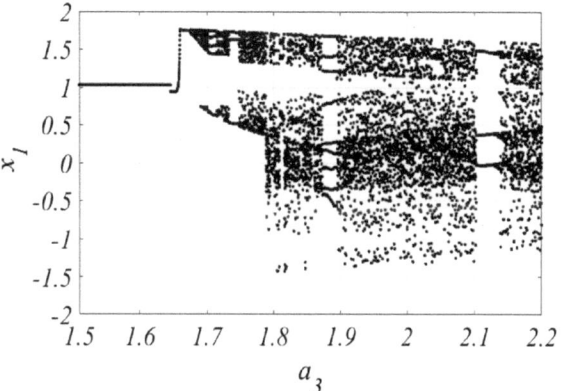

The numerical simulation of system (3.2) with the parameter values selected as $a_1 = 0.1$, $a_2 = 4$, $a_3 = 1.82$, $a_4 = 0.6$, $g = 10$, and $\tau = 1$ displays:

(i)   chaotic double-scroll attractor when $M = 1$ with nonlinear function (3.3);
(ii)  chaotic three-scroll attractor when $M = 1$ with nonlinear function (3.4);
(iii) chaotic four-scroll attractor when $M = 2$ with nonlinear function (3.3) as shown in Fig. 3.5a, b and c, respectively.

### 3.1.2 A New Nonautonomous Memristive Time-Delay System

A new system is proposed by introducing another state to the time-delay chaotic system (3.2) and driven by staircase function as follow:

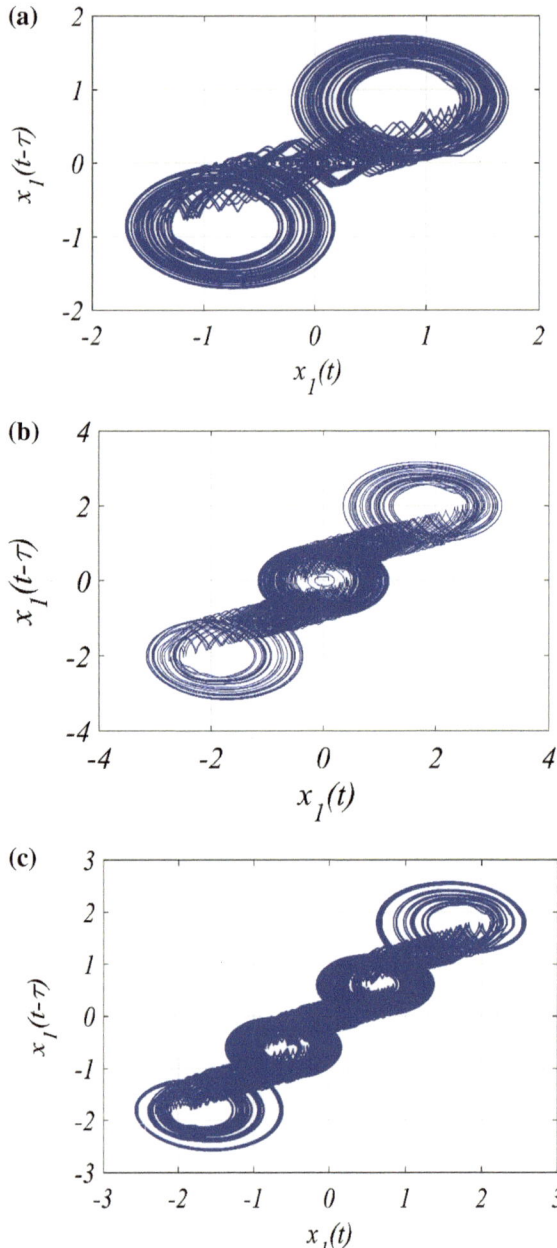

**Fig. 3.5** Numerical simulations of the n-scroll attractors, $\tau = 1$ s, $a_1 = 0.1$, $a_2 = 4$, $a_3 = 1.82$, $a_4 = 0.6$, $g = 10$, **a** two-scroll ($M = 1$, $r = 4$); **b** three-scroll ($M = 1$, $r = 6$); and **c** four-scroll ($M = 2$, $r = 4$). The initial conditions $(x_1(t),\ x_2(t)) = (0.1,\ 0.1)$

$$\begin{cases} \dot{x}_1 = F\left(x_{1\tau_1}\right) - a_1 h(x_1, x_3) \\ \dot{x}_2 = F\left(x_{2\tau_2}\right) + c\left(x_{1\tau_1} + P(u) - x_2\right) \\ \dot{x}_3 = x_1 - x_1 x_3 - a_2 x_3 \end{cases} \tag{3.6}$$

where $P(u)$ is a staircase function defined by:
for the $2N$-scroll attractors in the $x_2$-direction

$$P_1(u) = A \sum_{k=-(N-1)}^{N-1} \text{sgn}\left[u + l|k|\text{sgn}(k)\right] \tag{3.7}$$

for the $(2N + 1)$-scroll attractors in the $x_2$-direction

$$P_2(u) = A \sum_{k=1}^{N} \text{sgn}[u - (lk - 1)] + \text{sgn}[u + (lk - 1)] \tag{3.8}$$

where $u = \sin(\omega t)$, $c$, $A$, and $l$ are positive parameters, $\tau_1$ and $\tau_2$ are the time-delays, $F(x_{1\tau_1})$ and $F(x_{2\tau_2})$ are nonlinear functions, and $h(x, x_3)$ is the output of the memristive element as presented in (2.12).

The switching constant of staircase function affects the position of equilibrium points in space. The number of equilibrium points is increased, because of the multi-level of staircase function.

The new system (3.6) with (3.3), (3.4), (3.7), and (3.8) can be classified into several regions based on the characteristic regions of nonlinear functions. Figure 3.6b, d, and f shows the regions of characteristic functions of 4-, 9- and 16-scroll chaotic attractors, respectively, where $\pm D_i (i = 0, 1, 2, \ldots, M)$ denotes different regions and $\pm x_{ei}$, $\pm x'_{ei}$, $\pm x''_{ei}$ $(i = 0, 1, 2, \ldots, M)$ are the equilibrium points. For the four-scroll chaotic attractor, there are six equilibrium points $x_{e0}$, $+x_{e1}$, $-x_{e1}$ and $x'_{e0}$, $+x'_{e1}$, $-x'_{e1}$ as shown in Fig. 3.6b; $x_{e0}$ and $x'_{e0}$ are unstable for all $\tau$, and the stability of $\pm x_{e1}$ and $\pm x'_{e1}$ is dependent on the value of $\tau$. For the nine and sixteen scrolls, there are fifteen $\pm x_{ei}$, $\pm x'_{ei}$, $\pm x''_{ei}$ $(i = 0, 1, 2)$ and twenty-eight $\pm x_{ei}$, $\pm x'_{ei}$, $\pm x''_{ei}$ $(i = 0, 1, 2, 3)$ equilibrium points as shown in Fig. 3.6d, f, respectively. The saddle point is unstable and independent on the value of $\tau$, and the remaining equilibrium points are unstable saddle-focus, which generate the scrolls. It means the new system can increase the number of equilibrium points depending on the switching of staircase function, while the stability of equilibrium points stays in the same with system (3.2).

The numerical simulation of system (3.6) with the parameter values selected as $c = 5$, $a_1 = 0.1$, $a_2 = 4$, $a_3 = 1.82$, $a_4 = 0.6$, $g = 10$, $A = 2$, $l = 0.5$, and $\tau_1 = \tau_2 = 1$ s displays:

(i)   chaotic four-scroll attractor when $M = 1$, $r = 4$ and $N = 1$ with nonlinear functions (3.3) and (3.7), respectively;

(ii)  chaotic nine-scroll attractor when $M = 1$, $r = 6$ and $N = 1$ with nonlinear functions (3.4) and (3.8), respectively;

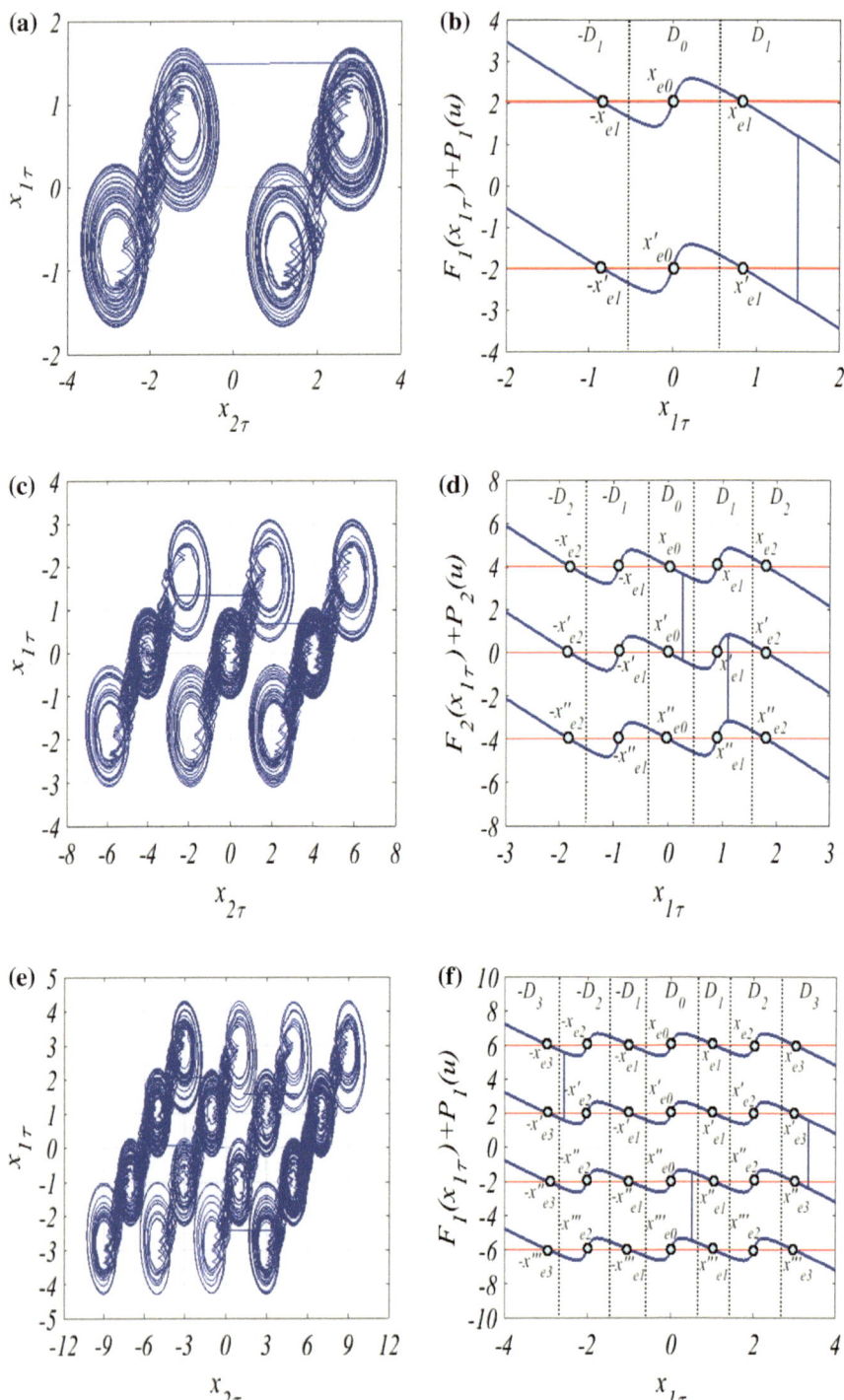

◀Fig. 3.6 **Fig. 3.6** Phase portraits and corresponding characteristics regions of 2D-grid scroll chaotic attractors, **a** and **b** 4-scroll attractor, **c** and **d** 9-scroll attractor, **e** and **f** 16-scroll attractor. The initial conditions $(x_1(t), x_2(t), x_3(t)) = (0.01, 0.1, -0.3)$

(iii)   chaotic 16-scroll attractor when $M = 2$, $r = 4$ and $N = 2$ with nonlinear functions (3.3) and (3.7), respectively, as shown in Fig. 3.6a, c, and e, respectively.

## 3.2 Circuit Realization of Memristive Time-Delay Systems

In this section, an electronic circuit is designed to realize the autonomous system (3.2) and nonautonomous system (3.6) with the nonlinear part (3.3) and (3.4) through the staircase functions (3.7) and (3.8) as shown in Fig. 3.7. The circuit is based on current feedback operational amplifier (CFOA-AD844). Figure 3.7 shows the circuit diagram. This circuit diagram includes seven parts; that are Part I: *switch* ($N_0$); Part II: *integrator* ($N_1$); Part III: *memristor* ($N_2$); Part IV: *summer* ($N_3$); Part V: *inverter* ($N_4$); Part VI: *time-delay circuit* ($N_5$); Part VII: *nonlinear function* ($N_6$); and Part VIII: *staircase function generator* ($N_7$). Note that the switch $N_0$ selecting if the system is autonomous or not as shown in Table 3.1:

The integration time constant of the integrators $N_1$ is determined by $R_1 C_1$ and $R_2 C_2$, and the range of frequency is changed by different values for $R_1$, $C_1$ and $R_2$, $C_2$. The model of the memristor element $N_2$ was proposed in the previous chapter. The configurations of CFOAs in part four $N_3$ are operated as summers for $X_1$ and $X_2$. The time-delay parts $N_5$ are implemented by using a cascade of five low-pass second-order Bessel filters as shown in Fig. 3.8a. Each one is characterized by the Sallen–Key topology with the following transfer function [4]:

$$H(s) = \frac{1}{1 + C_{10}(R_{20} + R_{30})s + C_{10}C_{20}R_{20}R_{30}s^2} \tag{3.9}$$

The values of filter components are chosen in order to realize a Bessel approximation with 3 dB frequency equal to $f_C = 1$ kHz. The time delay introduced by this filter in the band up to $f_c$ can be calculated as [4]:

$$\tau \simeq C_{10}(R_{20} + R_{30}) \tag{3.10}$$

There are some advantages of this filter make it's used. Firstly, the CFOA is configured as an amplifier, so the configuration shows the least dependence of filter performance on the CFOA. Another advantage of this configuration is that the ratio of the largest resistor value to the smallest resistor value and the ratio of the largest capacitor value to the smallest capacitor value (component spread) are low, which is

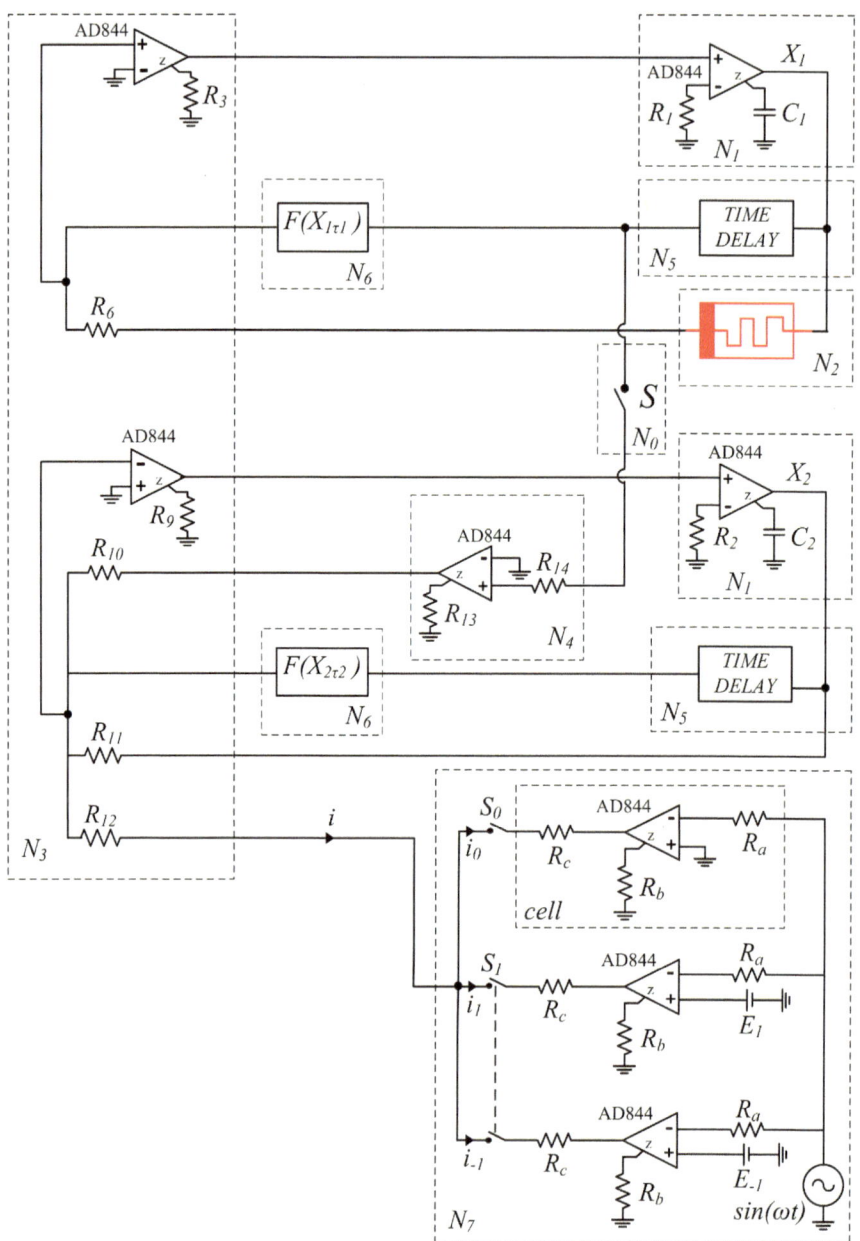

**Fig. 3.7** Circuit diagram for realizing $n$-scroll chaotic attractors with systems (3.2) and grid scroll system (3.6), and nonlinear functions expressed in the systems (3.3) and (3.4), with the staircase functions (3.7) and (3.8). The values of the components are selected as $R_1 = R_2 = R_c = 10$ k, $R_3 = R_6 = R_9 = 2$ k, $R_{10} = R_{11} = 0.4$ k, $R_{12} = 6$ k, $R_{13} = R_{14} = R_a = 1$ k, $R_b = 100$ k, $C_1 = C_2 = 100$ nF, $V_{cc} = \pm 12$ V, $V_{sat} \approx \pm 8.4$ V

**Table 3.1** ON–OFF switch linkage $S$ for selecting type of system

| State of switch $S$ | Classification of system |
| --- | --- |
| OFF | Autonomous |
| ON | Nonautonomous |

**(a)**

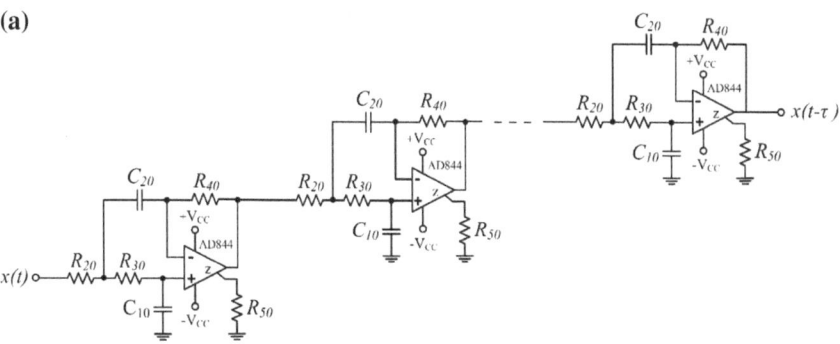

**(b)**

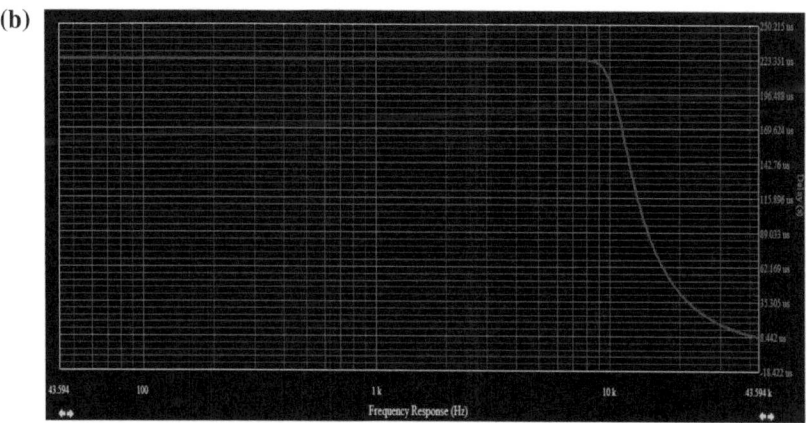

**Fig. 3.8** **a** Schematic of the Sallen–Key low-pass active filter including five parts in cascaded for implementing the time delay $\tau = 1$ ms. $R_{20} = R_{30} = 10$ k$\Omega$, $R_{40} = 1$ k$\Omega$, $R_{50} = 100$ k$\Omega$, $C_{10} = 10$ nF, $C_{20} = 22$ nF. **b** Group delay spectrum of the filter

good for manufacturability. The most important property is that the group delay of this filter is constant as shown in Fig. 3.8b [5].

The nonlinear function $F(X)$, $N_6$ in Fig. 3.7, is implemented by the circuitry of Fig. 3.9. Since the large signal behavior of a bipolar transistor differential cell can accurately be modeled using smooth inverse tangent function as shown in Fig. 3.9 (2N3904 transistor used to implement the nonlinear function), a block composed of properly biased alternating bipolar transistor differential cells is constructed to realize this nonlinearity. The usefulness of the differential pair stems from two key properties.

**Table 3.2** ON–OFF of linkage switches $S_{11} \sim S'_{1j}$ and number of scrolls

| State of linkage switches | | | | | | | Number of scrolls |
|---|---|---|---|---|---|---|---|
| $S_{11}$ | $S'_{11}$ | $S_{12}$ | $S'_{12}$ | $\cdots$ | $S_{1j}$ | $S'_{1j}$ | |
| OFF | ON | OFF | OFF | $\cdots$ | OFF | OFF | 2 |
| OFF | OFF | OFF | ON | $\cdots$ | OFF | OFF | 3 |
| ON | OFF | OFF | ON | $\cdots$ | OFF | OFF | 4 |
| $\vdots$ | $\vdots$ | $\vdots$ | $\vdots$ | $\cdots$ | $\vdots$ | $\vdots$ | $\vdots$ |
| OFF | ON | ON | OFF | $\cdots$ | OFF | ON | $n$ − odd |
| ON | OFF | ON | OFF | $\cdots$ | OFF | ON | $n$ − even |

First, cascades of differential pairs can be directly connected to one another without interstage coupling capacitors. Second, the differential pair is primarily sensitive to the difference between two input voltages, allowing a high degree of signals rejection common to both inputs. The circuit includes switches for selecting among different nonlinear function. Table 3.2 reports the states of the switches to implement the nonlinearity for the circuit to generate an attractor with $n$-scroll. For example, to generate attractor of 3-scroll, the linkage switches are fixed to $S_{11} = S'_{11} = S_{12} = 0$, $S'_{12} = 1$. The nonlinear device is described by the circuital equations as follows:

- for even number of scrolls:

$$F(X) = -\frac{R_3}{R_5}X + \sum_{p=-(m-1)}^{m-1} \frac{R_3}{R_4} \tan^{-1}\left(\frac{R_7}{R_8}X + V_p\right) \tag{3.11}$$

- for odd number of scrolls:

$$F(X) = -\frac{R_3}{R_5}X + \sum_{p=-(m-1)}^{m-1} \frac{R_3}{R_4}\left(\tan^{-1}\left(\frac{R_7}{R_8}X + V_p\right) + \tan^{-1}\left(\frac{R_7}{R_8}X - V_p\right)\right) \tag{3.12}$$

where $m = (1, 2, 3 \ldots)$, and $V_p$ is the base control voltage ($p = 1, 2, \ldots, j$).

The staircase function generator, $N_7$, can be designed with CFOA. Each cell in this part shows the shifted voltage sign function. By connecting the cells in parallel, the staircase function is implemented. The total current of the staircase function with an even number of scrolls can be obtained as follows:

$$i = S_0 i_0 + S_1(i_1 + i_{-1})$$

$$= P_1(\sin(\omega t))$$

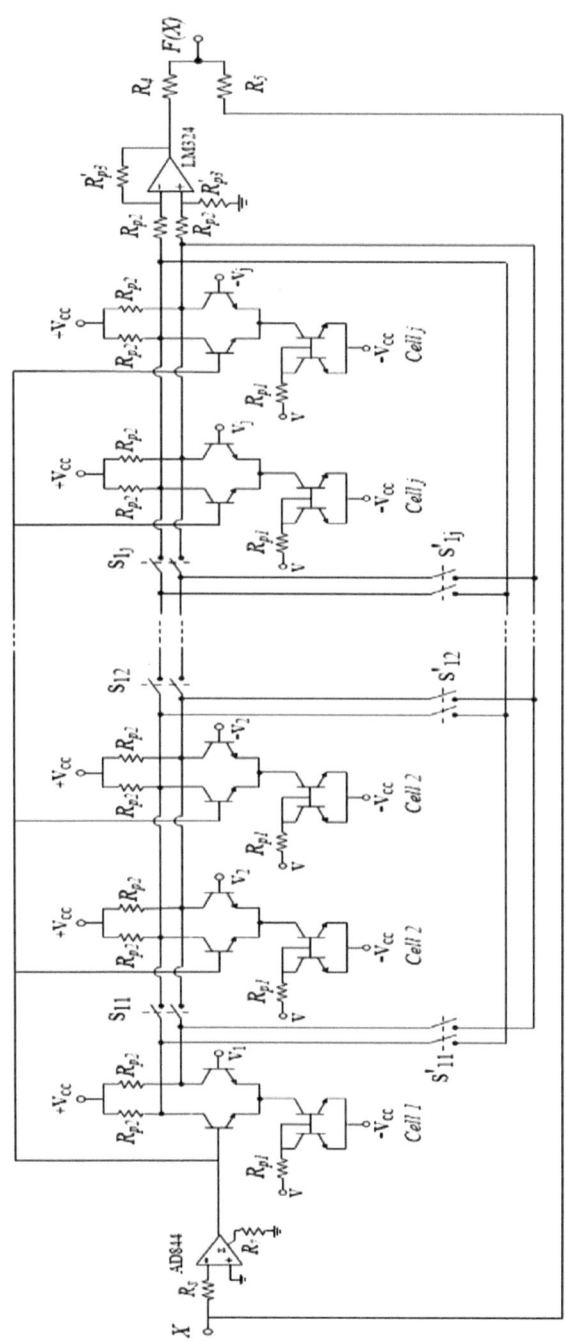

**Fig. 3.9** Schematic of the nonlinear function: $R_{p1} = 1.2$ k, $R_{p2} = 1$ k, $R_{p3} = 1$ k, $R'_{p3} = 3$ k, $R_4 = 3$ k, $R_5 = 1.25$ k, $R_7 = 10$ k, $R_8 = 1$ k; $V_{cc}(\pm 12$ V); $V = -10.3$ V; (three-scroll $V_2 = 3$V; $V_2 = -3$ V); (four-scroll $V_1 = 0$ V; $V_2 = 3.4$ V; $-V_2 = -3.4$ V)

$$i = \sum_{j=-N}^{N} -\frac{|V_{\text{sat}}|}{R_c} \text{sgn}\big[\sin(\omega t) - |j|E_j\big] \tag{3.13}$$

where $N = 1$.

Similarly, from the circuit diagram shown in Fig. 3.7, one can get the total current with an odd number of scrolls as follows:

$$i = S_1(i_1 + i_{-1})$$
$$= P_2(\sin(\omega t))$$

$$i = \sum_{\substack{j=-N \\ j \neq 0}}^{N} -\frac{|V_{\text{sat}}|}{R_c} \text{sgn}\big[\sin(\omega t) - E_j\big] \tag{3.14}$$

where $N = 1$, $E_j$ is switching points, and $V_{\text{sat}}$ is the saturated value of the AD844 device.

Referring to the circuit theory and the circuit diagram shown in Fig. 3.7, one can write finally the following equation:

**Case 1**: The switch $S$ is OFF (autonomous system)

$$\frac{dX_1(t)}{dt} = \frac{1}{R_1 C_1}\left(F\big(X_{1\tau_1}\big) - \frac{R_3}{R_6}h(X_1, X_2)\right) \tag{3.15}$$

**Case 2**: The switch $S$ is ON (nonautonomous system)

$$\begin{cases} \frac{dx_1(t)}{dt} = \frac{1}{R_1 C_1}\left(F\big(X_{1\tau_1}\big) - \frac{R_3}{R_6}h(X_1, X_2)\right) \\ \frac{dx_2(t)}{dt} = \frac{1}{R_2 C_2}\left(F\big(X_{2\tau_2}\big) + \frac{R_9 R_{13}}{R_{10} R_{14}}X_{1\tau_1} - \frac{R_9}{R_{12}}P(\sin(\omega t)) - \frac{R_9}{R_{11}}X_2\right) \end{cases} \tag{3.16}$$

To generate multi-scroll from the circuit diagram shown in Fig. 3.7, the switch $S$ is set to OFF. Figure 3.10 shows the phase portrait in the plane $X_1 - X_{1\tau}$ and corresponding power spectrum of the signal of $X_1(t)$. For double-scroll attractor, the base control voltage is fixed to $V_1 = 0$ V and the linkage switches $(S_{11}, S'_{11})$ are set to (OFF,ON). The observed results are shown in Fig. 3.10a. For generating three-scroll attractor, the base control voltage is set to $V_2 = 2$ V, and the linkage switches $S_{11}, S'_{11}$, and $S_{12}$ are OFF, and $S'_{12}$ is ON. The captured results are shown in Fig. 3.10b. The four-scroll attractor can be generated by fixing the base control voltages as $V_1 = 0$ V, $V_2 = 3.4$ V; also, the linkage switches $S'_{11}$ and $S_{12}$ are set to OFF, while $S_{11}$ and $S'_{12}$ are ON. The PSpice results as shown in Fig. 3.10c.

According to circuit diagram shown in Fig. 3.7, Tables 3.2 and 3.3, one can generate $2D$ ($n \times m -$ scroll) chaotic attractors with different values of $N$ and $M$, as shown in Fig. 3.11.

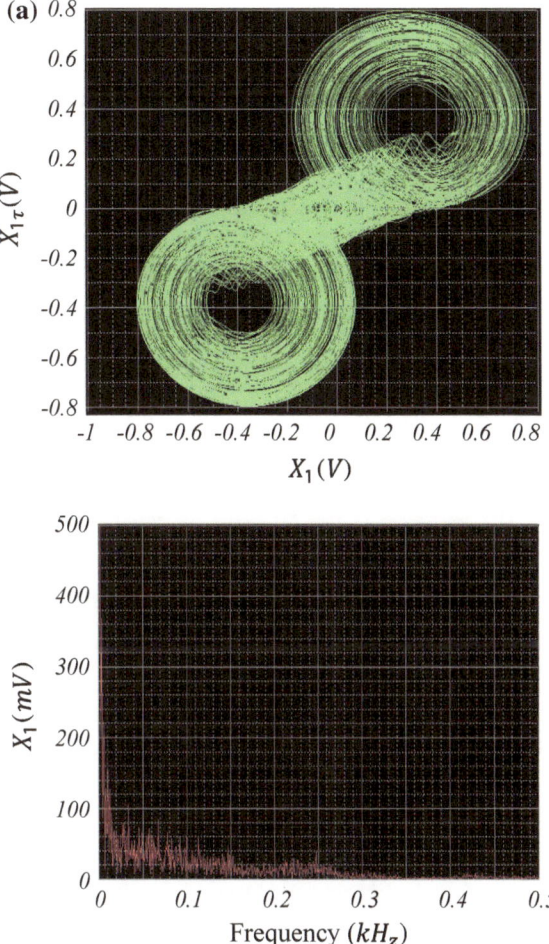

**Fig. 3.10** PSpice simulation results. Phase portrait and corresponding power spectrum for autonomous system (3.2). **a** Two-scroll attractor, **b** three-scroll attractor, and **c** four-scroll attractor are present in $X_1 - X_{1\tau}$ plane

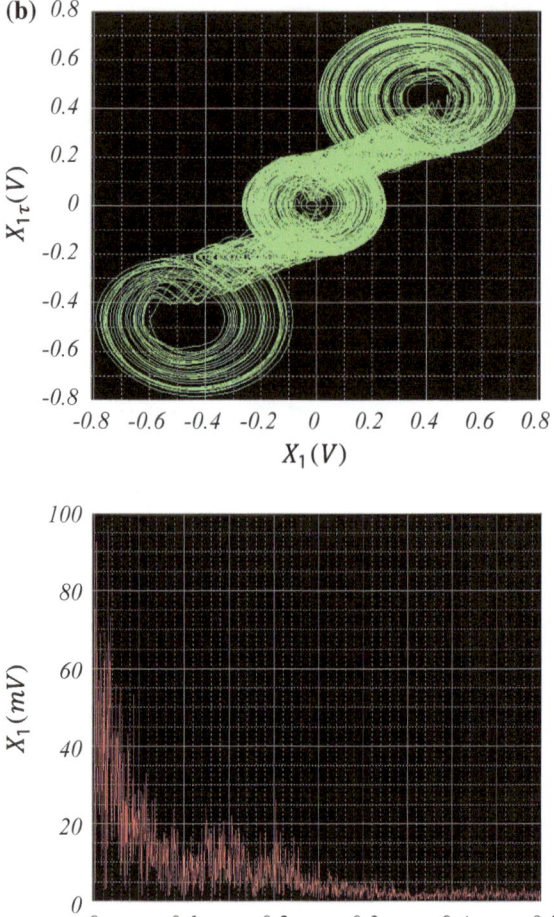

**Fig. 3.10** (continued)

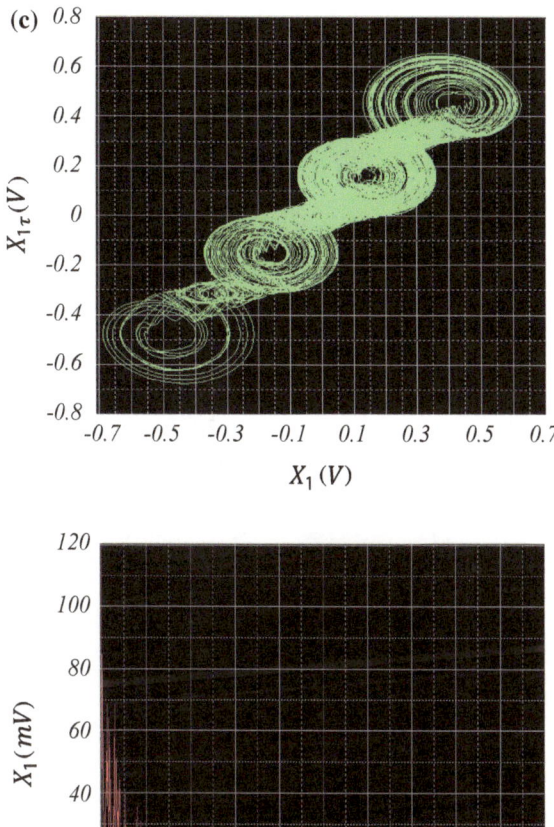

Fig. 3.10 (continued)

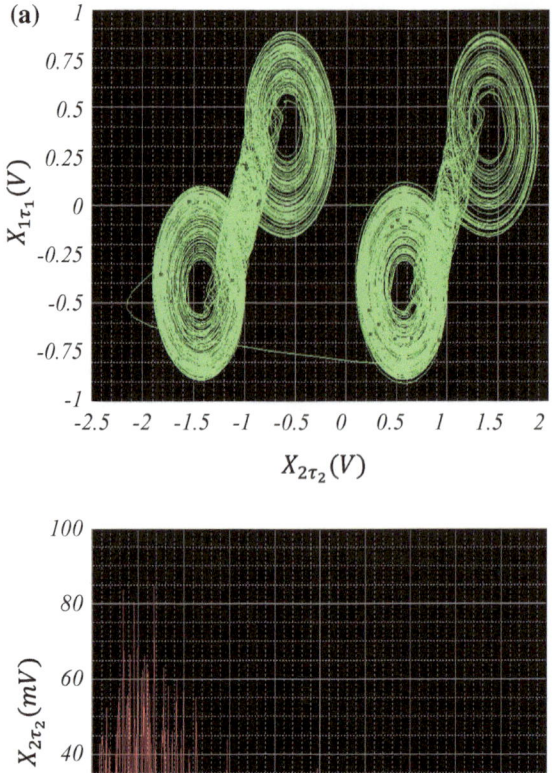

**Fig. 3.11** PSpice simulation results. Phase portrait and corresponding power spectrum for 2D-grid scroll nonautonomous system (3.6). **a** $2 \times 2$-scroll attractor, **b** $3 \times 3$-scroll attractor, and **c** $4 \times 4$-scroll attractor are present in $X_{2\tau_2} - X_{1\tau_1}$ plane

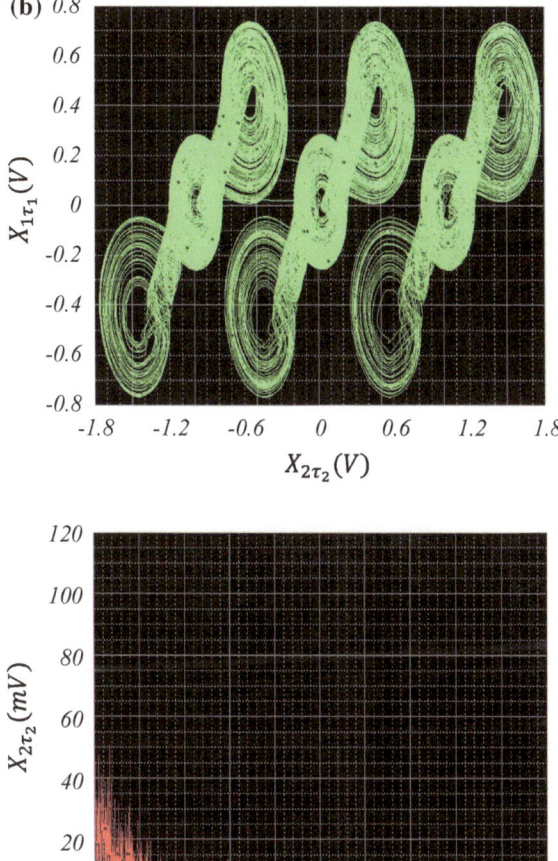

**Fig. 3.11** (continued)

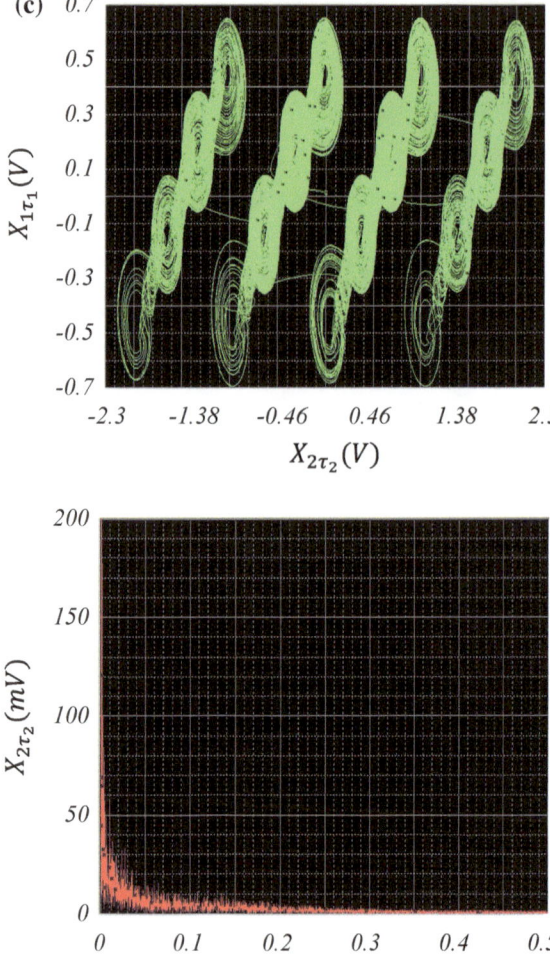

**Fig. 3.11** (continued)

**Table 3.3** ON–OFF of linkage switches $S_0$, $S_1$ and number of scrolls

| State of linkage switches | | Number of scrolls |
|---|---|---|
| $S_0$ | $S_1$ | |
| ON | OFF | 2 |
| OFF | ON | 3 |
| ON | OFF | 4 |

Suppose that the switch $S$ is ON. The state of shifts is selected as $E_1 = 2.5$ V and $E_{-1} = -2.5$ V. By selecting the $(S_{11}, S_1)$ and $(S'_{11}, S_0)$ as (OFF and ON), then, the circuit diagram can generate $2D$ ($2 \times 2$ − scroll) chaotic attractor as shown in Fig. 3.11a. The circuit diagram can generate $2D$ ($3 \times 3$ − scroll) chaotic attractor by setting the switch $S$ is ON, $(S_{11}, S'_{11}, S_{12}, S_0)$ and $(S'_{12}, S_1)$ are (OFF and ON) as shown in Fig. 3.11b. For generating $2D$ ($4 \times 4$ − scroll), chaotic attractor must select the switches as follow: $S$ is ON, $(S'_{11}, S_{12}, S_1)$ and $(S_{11}, S'_{12}, S_0)$ are (OFF and ON) as shown in Fig. 3.11c.

## 3.3 A New Two Memristor-Based Chaotic System

According to the model in [6] and based on the memristor model (2.12), a new memristive system is introduced. The new system has two memristors. A chaotic circuit with two memristors is proposed and implemented. The new circuit dynamic consists of five dimensions. The mathematical model of the proposed system to generate nonlinear dynamics is as follows:

$$\begin{cases} \dot{x}_1(t) = \sigma x_2 \\ \dot{x}_2(t) = \beta h(x_1, x_4) + \alpha x_3 - \gamma x_2 \\ \dot{x}_3(t) = 1 - \alpha h(x_3, x_5) - x_1 \\ \dot{x}_4(t) = x_1 - x_1 x_4 - \delta x_4 \\ \dot{x}_5(t) = x_3 - x_3 x_5 - \delta x_5 \end{cases} \tag{3.17}$$

where $x_1, x_2, x_3, x_4$, and $x_5$ are five independent dynamical variables, while $\sigma, \beta, \alpha, \gamma$, and $\delta$ are system parameters, and $h(x_3, x_5)$ and $h(x_1, x_4)$ represent the output of the memristor elements as presented in (2.12).

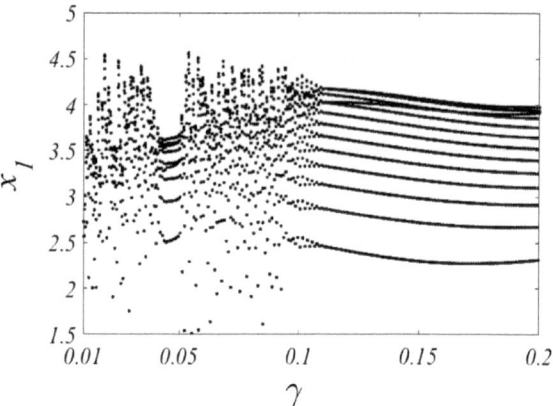

**Fig. 3.12** Bifurcation diagram for the proposed memristive system (3.17) when $\sigma = 2$, $\beta = 0.5$, $\alpha = 0.1$, $\delta = 4$, the initial conditions $(x_1(0), x_2(0), x_3(0), x_4(0), x_5(0)) = (0.01, 0.1, 0.01, 0, 0)$ and $\gamma$ as a varying parameter

The system (3.17) displays chaotic attractor with no equilibrium points. When the parameters of system (3.17) are set as: $\sigma = 2$, $\beta = 0.5$, $\alpha = 0.1$, $\delta = 4$ and the initial conditions are $(x_1(0), x_2(0), x_3(0), x_4(0), x_5(0)) = (0.01, 0.1, 0.01, 0, 0)$, while the parameter $\gamma$ is varied. The maxima of the state variable $x_1$ is plotted with $\gamma$. The bifurcation diagram is formed as shown in Fig. 3.12. For the selected parameters, the phase portraits of system are shown in Fig. 3.13a, b, and c.

### 3.3.1 Circuitry Realization

In this section, an electronic circuit is designed to realize the proposed chaotic system (3.17) with two memristors as shown in Fig. 3.14. For this purpose, the current feedback operational amplifier is employed. Figure 3.14 shows the circuit scheme. It includes five parts; that are Part I: *integrator* ($N_0$); Part II: *memristor* ($N_1$); Part III: *inverter* ($N_2$); Part IV: *summer* ($N_3$); and Part V: *gain* ($N_4$). The integration time constant of the integrators $N_0$ is determined by $R_1C_1$, $R_2C_2$, and $R_3C_3$. This circuit operated with two memristor elements $N_1$.

Each inverter $N_2$ multiplies the input's signal by a factor equal to negative gain of it ($-R_8/R_9$, $-R_{10}/R_{11}$ and $-R_{12}/R_{13}$). The summer circuits $N_3$, are using for summing three terms needed for the $X_2$ and $X_3$ variables. The first summer (on the left) is summing ($\frac{-R_8}{R_9}X_3$, $\frac{-R_{10}}{R_{11}}h(X_1, X_4)$, $X_2$), while the second summer (on the right) is summing ($X_1$, $h(X_3, X_5)$, $E$). The final part $N_4$ is used to multiply the $X_2$ variable by the gain ($-R_{14}/R_{15}$).

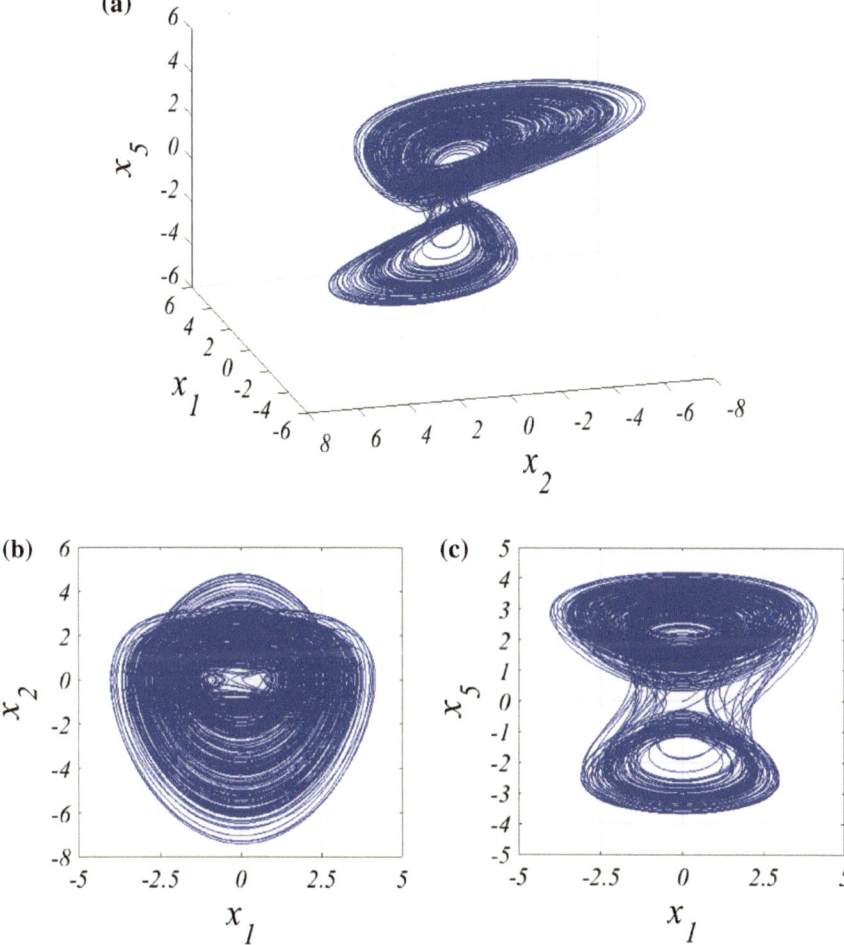

**Fig. 3.13** Phase portraits of the new memristive chaotic system (3.17) for $\sigma = 2$, $\beta = 0.5$, $\alpha = 0.1$, $\delta = 4$, and $\gamma = 0.02$, the initial conditions $(x_1(0), x_2(0), x_3(0), x_4(0), x_5(0)) = (0.01, 0.1, 0.01, 0, 0)$ in **a** 3D view in $x_1 - x_2 - x_5$ space, **b** $x_1 - x_2$ plane, and **c** $x_1 - x_5$ plane

By applying Kirchhoff's circuit laws, the corresponding circuital equations of the circuit can be written as follows:

$$\frac{dX_1(t)}{dt} = \frac{1}{R_1 C_1} \left( \frac{R_{12} R_{14}}{R_{13} R_{15}} X_2 \right)$$

$$\frac{dX_2(t)}{dt} = \frac{1}{R_2 C_2} \left( \frac{R_8 R_4}{R_9 R_7} X_3 + \frac{R_{10} R_4}{R_{11} R_6} h(X_1, X_4) - \frac{R_4}{R_5} X_2 \right)$$

$$\frac{dX_3(t)}{dt} = \frac{1}{R_3 C_3} \left( -\frac{R_{16}}{R_{18}} E - \frac{R_{16}}{R_{19}} h(X_3, X_5) - \frac{R_{16}}{R_{17}} X_1 \right) \qquad (3.18)$$

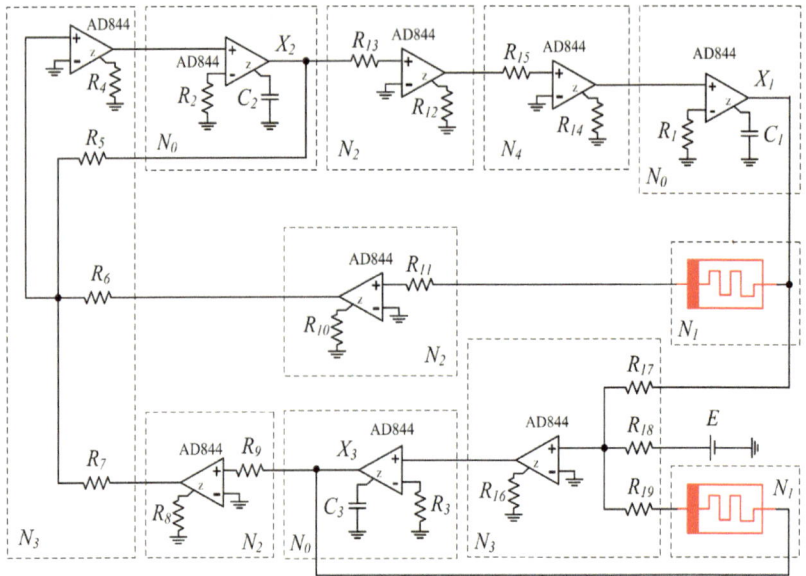

**Fig. 3.14** Circuit diagram for realizing two memristor-based chaotic systems (3.17): The values of the components are selected as $R_1 = R_2 = R_3 = 10$ k, $R_8 = R_9 = R_{10} = R_{11} = R_{12} = R_{13} = R_{15} = R_{16} = R_{17} = R_{18} = R_{19} = 1$ k, $R_4 = R_6 = R_{14} = 2$ k, $R_5 = 100$ k, $R_7 = 20$ k, $C_1 = C_2 = C_3 = 100$ nF

where $X_1$, $X_2$, and $X_3$ are voltages across the capacitors $C_1$, $C_2$, and $C_3$. $X_4$ and $X_5$ are the internal state of the memristors. The equations of these states were not appearing in Eq. (3.18) and not obvious in the circuit diagram of the system because it is designed and implemented in chapter two.

The circuit consists of common off-the-shelf discrete components such as capacitors, resistors, and current feedback operational amplifiers (CFOA)-(AD844). The power supplies are fixed to $\pm 12$ V and $E = -1$ V. Equation (3.18) matches Eq. (3.17) with $\sigma = \frac{R_{12}R_{14}}{R_{13}R_{15}}$, $\beta = \frac{R_{10}R_4}{R_{11}R_6}$, $\alpha = \frac{R_8R_4}{R_9R_7}$, and $\gamma = \frac{R_4}{R_5}$.

The circuit diagram depicted in Fig. 3.15 is successfully implemented using the PSpice software; the practical results are shown in chaotic attractors and the corresponding power spectrums as depicted in Fig. 3.15a, b, c, and d.

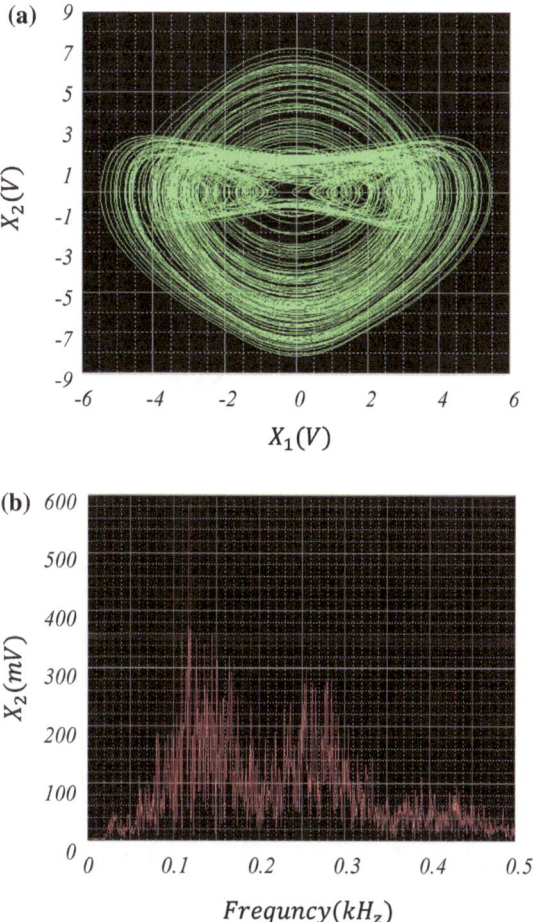

**Fig. 3.15** Phase portrait and corresponding power spectrum for chaotic system (3.17). **a** $X_1 - X_2$ plane, **b** power spectrum of $X_2$ state, **c** $X_1 - X_5$ plane, and **d** power spectrum of $X_5$ state

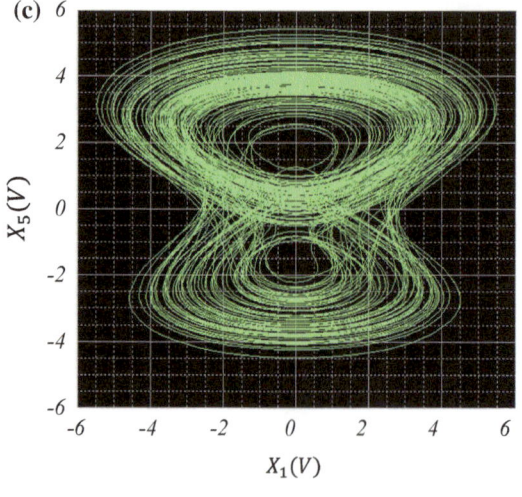

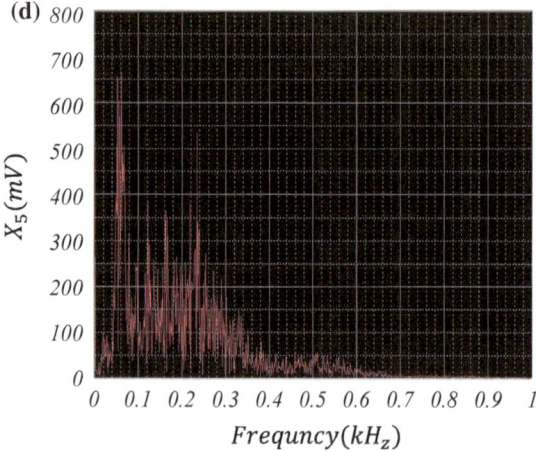

**Fig. 3.15**  (continued)

# References

1. A. T. Azar, S. Vaidyanathan, *Chaos Modeling and Control Systems Design*, vol. 581. (Springer International Publishing Switzerland, 2015) ISBN 978-3-319-13131-3
2. P. Amil, C. Cabeza, A.C. Marti, Exact discrete-time implementation of the Mackey-Glass delayed model. IEEE Trans. Circuits Syst. II Express Briefs **62**(7), 681–685 (2015)
3. M. Asadi, F. Salehi, S. Mohyud-Din, M. Hosseini, Modified homotopy perturbation method for stiff delay differential equations (DDEs). Int. J. Phys. Sci. **7**(7), 1025–1034 (2012)
4. A. Buscarino, L. Fortuna, M. Frasca, G. Sciuto, Design of time-delay chaotic electronic circuits. IEEE Trans. Circuits Syst. I Regul. Pap. **58**(8), 1888–1896 (2011)
5. R.P. Sallen, E.L. Key, A practical method of designing RC active filters. IRE Trans. Circuit Theory **2**(1), 74–85 (1955)
6. A. Buscarino, C. Corradino, L. Fortuna, M. Frasca, J.C. Sprott, Nonideal behavior of analog multipliers for chaos generation. IEEE Trans. Circuits Syst. II Express Briefs **63**(4), 396–400 (2016)

# Chapter 4
# Synchronization of Memristive Electronic Circuits

The synchronization between nonlinear electronic circuits is not reachable, because these circuits are extremely sensitive to initial conditions. Pecora and Carroll (PC), in 1990 was synchronized two chaotic circuits with different initial conditions. The synchronization phenomenon occurs when the trajectories of the two circuits converge to the same values, and each one will stay on track with the other. In master–slave configuration, the synchronization is achieved when the master circuit can drive the slave circuit. The master output signal controls the dynamics of the slave circuit, so the dynamics of a slave circuit track the master's dynamics asymptotically [1, 2].

This chapter is organized as follows. The chaos synchronization methods are addressed which are used in the nonlinear circuits. Two methods are selected and applied on the memristive time-delay chaotic circuit; the first one is PC synchronization, and the other is feedback control synchronization.

## 4.1 Chaos Synchronization

The techniques of synchronization which are suitable for standard communications systems can be generally classified as follows [3, 4]:

(1) *Method I*: Master–slave system (when the system parameters are known).
(2) *Method II*: Adaptive control system (when the receiver system parameters are unknown).

In this book, we will consider known system parameters (with two cases match and mismatch). The synchronization methods used are PC and feedback control.

© The Author(s), under exclusive license to Springer Nature Switzerland AG 2019
F. Rahma and S. Muneam, *Memristive Nonlinear Electronic Circuits*,
SpringerBriefs in Nonlinear Circuits, https://doi.org/10.1007/978-3-030-11921-8_4

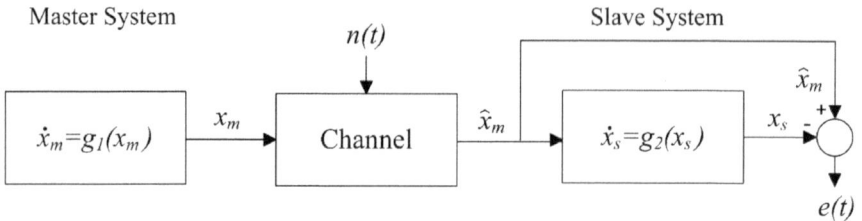

**Fig. 4.1** Scheme of PC synchronization

## 4.1.1 Pecora and Carroll Method

Pecora and Carroll (PC) synchronization is a pioneering scheme which is the earliest form of synchronization which was introduced for synchronizing chaotic circuits with different initial conditions, in which the transmitter (master) system drives the receiver (slave) system [1].

The master chaotic system can be given by:

$$\dot{x}_m(t) = g_1(x_m) \tag{4.1}$$

The slave chaotic system is described by:

$$\dot{x}_s(t) = g_2(x_s) \tag{4.2}$$

where $g_1$ and $g_2$ are the nonlinear functions of master and slave systems, respectively. The PC synchronization is satisfied by achieving the following condition:

$$\lim_{t \to \infty} ||x_m(t) - x_s(t)|| = 0 \tag{4.3}$$

where $||.||$ denotes the Euclidean norm and m and s indicate the master and the slave, respectively.

Figure 4.1 illustrates the scheme of PC synchronization. It shows the variable $x_m$ and how to transmit it through the channel to the receiver. The error $e(t)$ can be obtained by subtracting the output of receiver $x_s$ from $\hat{x}_m$ ($\hat{x}_m$ is the transmitted signal with noise).

The PC synchronization method is applied between the two memristive time-delay chaotic systems (master–slave) which were defined in (3.2) as follows:

The master system is expressed as:

$$\begin{cases} \dot{x}_{1m} = F\left((x_{1m})_{\tau_m}\right) - a_{1m}h(x_{1m}, x_{2m}) \\ \dot{x}_{2m} = x_{1m} - x_{1m}x_{2m} - a_{2m}x_{2m} \end{cases} \tag{4.4}$$

The slave system is given by:

$$\begin{cases} \dot{x}_{1s} = F\left((\hat{x}_{1m})_{\tau_s}\right) - a_{1s}h(x_{1s}, x_{2s}) \\ \dot{x}_{2s} = x_{1s} - x_{1s}x_{2s} - a_{2s}x_{2s} \end{cases} \tag{4.5}$$

where

$$F(x) = -a_{3m,s}x + a_{4m,s} \sum_{k=-(M-1)}^{M-1} \tan^{-1}(gx + kr) \tag{4.6}$$

The synchronization error $e(t)$ is given by:

$$e(t) = \hat{x}_{1m}(t) - x_{1s}(t) \tag{4.7}$$

The complete synchronization can be achieved by coupling of the master (4.4) and the slave (4.5) with match parameters and the $\lim_{t \to \infty} \|e(t)\| = 0$. In the mismatch parameters, $\lim_{t \to \infty} \|e(t)\| \neq 0$, $a_{1m} \neq a_{1s}$, $a_{2m} \neq a_{2s}$, $a_{3m} \neq a_{3s}$, and $a_{4m} \neq a_{4s}$ the synchronization does not occur. In the two cases (match and mismatch parameters), the delays and the initial conditions of the master and slave systems are selected as: $\tau_{m,s} = 1s$, $(x_{1m}(0), x_{2m}(0)) = (0.1, -0.1)$ and $(x_{1s}(0), x_{2s}(0)) = (0.3, 2)$. The Simulink model shown in Fig. 4.2 illustrates the master–slave configuration based on PC synchronization. In the case of match parameters, the synchronization is achieved between transmitted signal $\hat{x}_{1m}$ and receiver signal $x_{1s}$ as shown in Fig. 4.3a. It is clearly noticed that from the time series and synchronization error as shown in Fig. 4.3b, c, respectively, the synchronization begins after 0.5s. The systems parameters are selected as follows: $a_{1m} = a_{1s} = 0.1$, $a_{2m} = a_{2s} = 4$, $a_{3m} = a_{3s} = 1.82$ and $a_{4m} = a_{4s} = 0.6$.

With mismatched parameters of master and slave systems, the PC synchronization is not achieved as shown in Fig. 4.4. The parameters of two systems are selected as follows: $a_{1m} = 0.1$, $a_{1s} = 0.3$, $a_{2m} = 4$, $a_{2s} = 5$, $a_{3m} = 1.82$ $a_{3s} = 1.8$ and $a_{4m} = 0.6$, $a_{4s} = 0.5$. Notice that the error of synchronization does not converge to zero.

The PC synchronization method has been tested on the real channel (noisy channel). The real channel is subjected to additive white Gaussian noise (AWGN). The noise $n(t)$ has been added to the transmitted signal $x_{1m}(t)$ as follows:

$$\hat{x}_{1m}(t) = x_{1m}(t) + n(t) \tag{4.8}$$

The synchronization is lost when SNR $= 5$ dB as shown in Fig. 4.5a. By increasing the signal-to-noise ratio, SNR $= 10$ dB, SNR $= 15$ dB, and SNR $= 20$ dB, the synchronization error decreased as shown in Fig. 4.5b–e.

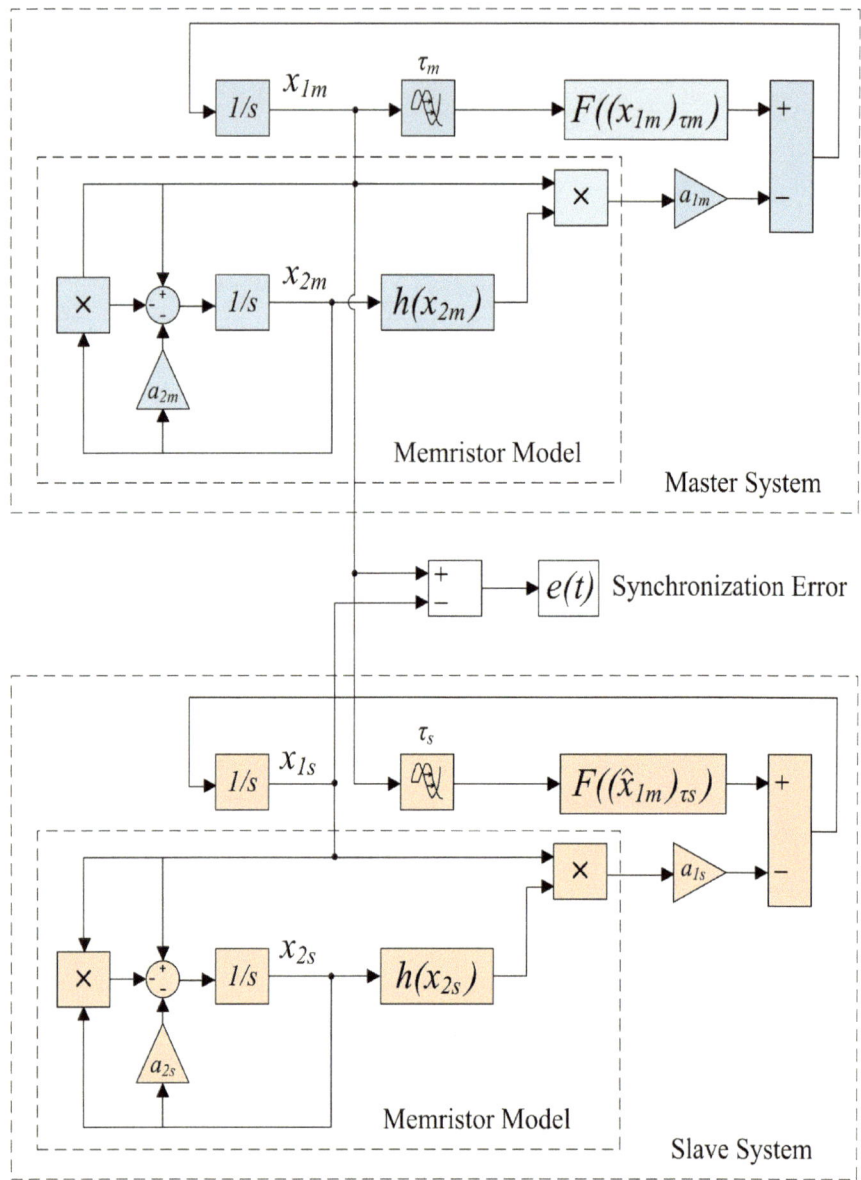

**Fig. 4.2** Simulink master–slave model PC synchronization

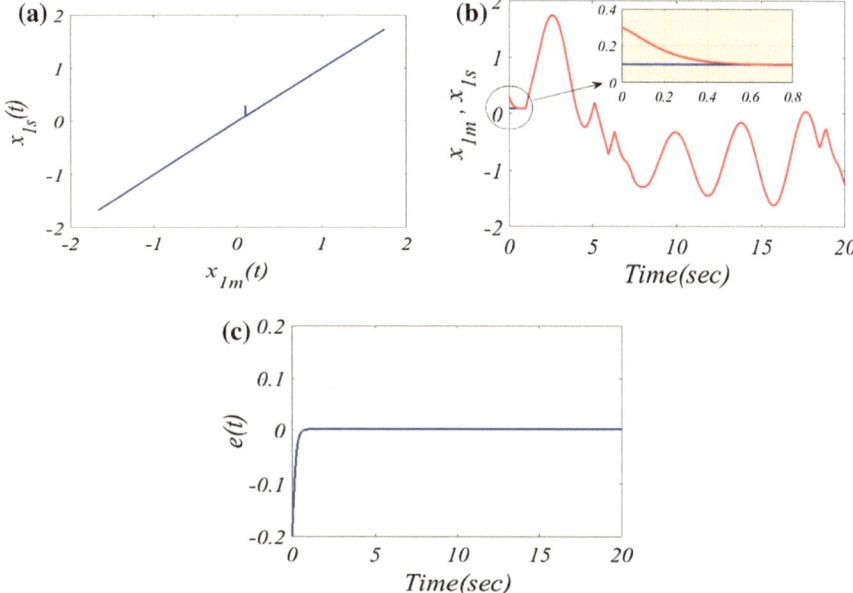

**Fig. 4.3** PC synchronization of master–slave configuration with match parameters and noiseless channel: **a** $x_{1m}(t)$ versus $x_{1s}(t)$, **b** time series of $x_{1m}(t)$ (blue) and $x_{1s}(t)$ (red), and **c** synchronization error

## 4.1.2 Synchronization via Control

The PC synchronization cannot be used in the communication systems where the parameters of transmitter and receiver are mismatched; also, the synchronization fails at the noisy channel. The feedback control method has been proposed to overcome these drawbacks.

The complete synchronization can be achieved between two memristive time-delay chaotic circuits (master–slave), while the parameters of two systems are not matched (the mismatch parameters are unavoidable in a real situation). The error of synchronization will not converge to zero, so this type is known as quasi-synchronization [5, 6]. Feedback control method can be used in the synchronization of two memristive time-delay chaotic circuits. The scheme of feedback control method shown in Fig. 4.6 is designed to perform the synchronization between two memristive time-delay chaotic systems. In this scheme, the slave system can be modified by adding the result of subtraction between the received signal $\hat{x}_{1m}$ and observed signal $x_{1s}$.

The master system is expressed as:

$$\begin{cases} \dot{x}_{1m} = F_m\big((x_{1m})_{\tau_m}\big) - a_{1m}h(x_{1m}, x_{2m}) \\ \dot{x}_{2m} = x_{1m} - x_{1m}x_{2m} - a_{2m}x_{2m} \end{cases} \tag{4.9}$$

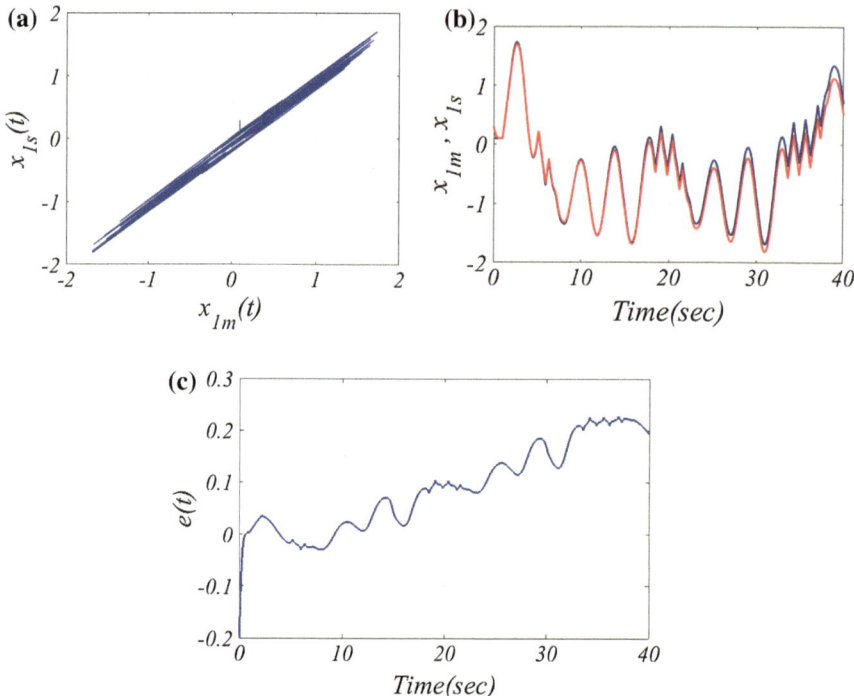

**Fig. 4.4**  PC synchronization of master–slave configuration with mismatch parameters and noiseless channel: **a** $x_{1m}(t)$ versus $x_{1s}(t)$, **b** time series of $x_{1m}(t)$ (blue) and $x_{1s}(t)$ (red), and **c** synchronization error

The slave system is given as follow:

$$\begin{cases} \dot{x}_{1s} = F_s\left(\left(\hat{x}_{1m}\right)_{\tau_s}\right) - a_{1s}h(x_{1s}, x_{2s}) + u(t) \\ \dot{x}_{2s} = x_{1s} - x_{1s}x_{2s} - a_{2s}x_{2s} \end{cases} \tag{4.10}$$

The control signal is given by:

$$u(t) = C_0\left(\hat{x}_{1m}(t) - x_{1s}(t)\right) = C_0 e(t) \tag{4.11}$$

where $C_0$ is feedback gain. It can be selected to reduce noise impact and achieve synchronization rapidly.

The Simulink model shown in Fig. 4.7 of the master–slave configuration is based on feedback control. In the case of match parameters, the synchronization is achieved between received signal $\hat{x}_{1m}$ and receiver signal $x_{1s}$ as depicted in Fig. 4.8. The parameters of two systems are selected as follows: $a_{1m} = a_{1s} = 0.1, a_{2m} = a_{2s} = 4, a_{3m} = a_{3s} = 1.82$, and $a_{4m} = a_{4s} = 0.6$. With mismatched parameters, the feedback control is conducted to achieve the quasi-synchronization between the received signal $\hat{x}_{1m}$

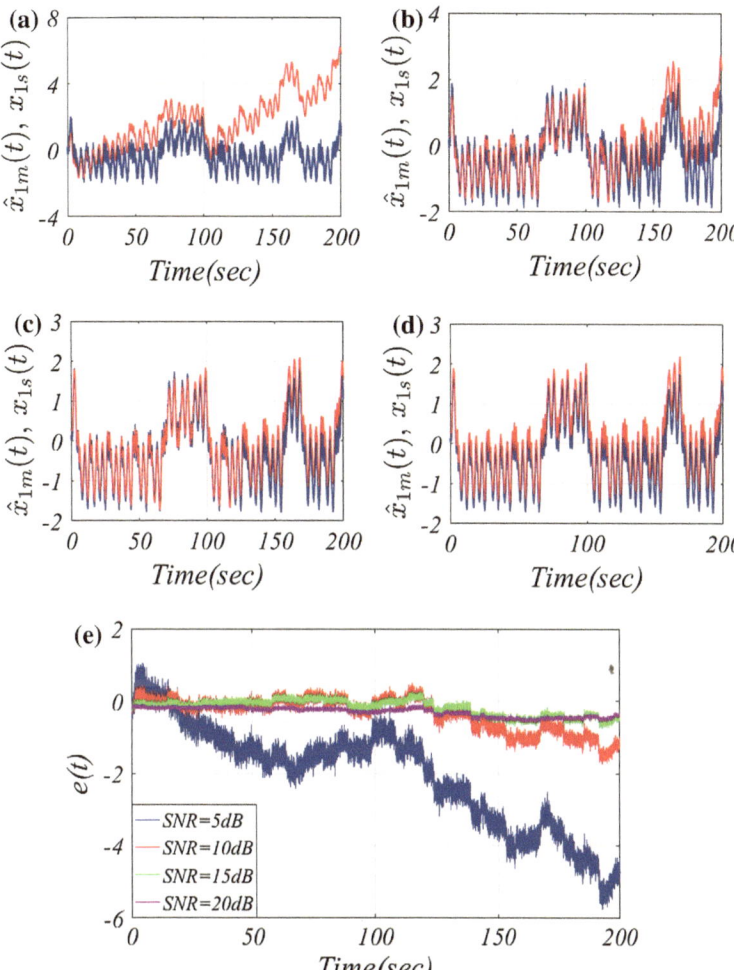

**Fig. 4.5**  Master (blue) and the slave (red) time series of the memristive time-delay paired systems in a PC synchronization with match parameters and noisy channel. **a** SNR = 5 dB, **b** SNR = 10 dB, **c** SNR = 15 dB, **d** SNR = 20 dB, and **e** synchronization errors

and the receiver signal $x_{1s}$ as shown in Fig. 4.9. The synchronization error oscillates about zero, and quality of synchronization depends on the feedback gain $C_0$. In this case, parameters of the two systems can be selected as follows: $a_{1m} = 0.1$, $a_{1s} = 0.3$, $a_{2m} = 4$, $a_{2s} = 5$, $a_{3m} = 1.82$, $a_{3s} = 1.8$ and $a_{4m} = 0.6$, $a_{4s} = 0.5$.

For real channel, we noticed the feedback control method is achieved quasi-synchronization in case of match parameters and the coupling strength $C_0 = 1$. With the presence of the noise, the error does not converge to zero. By increasing the signal to noise ratio from SNR = 5 dB to SNR = 20 dB, we notice that the error of synchronization is decreased as shown in Fig. 4.10 decreased when increasing SNR.

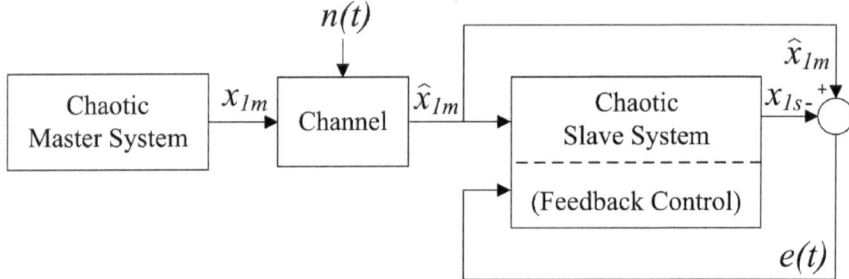

**Fig. 4.6** Synchronization scheme via feedback control

## 4.2  Circuit Realization

In this section, the electronic circuits of the two methods of synchronization have been designed and implemented. The circuits based on current feedback operational amplifiers (CFOA). The circuits consist of popular off-the-shelf discrete components such as capacitors, resistors, multipliers (AD633), and operational amplifiers (AD844).

### 4.2.1  PC Synchronization Electronic Circuit

Figure 4.11 shows the circuit diagram of PC synchronization. This circuit diagram includes several parts in master and slave circuits; that are Part I: *memristor* ($N_0$); Part II: *time-delay circuit* ($N_1$); Part III: *nonlinear function* ($N_2$).

The basic blocks had been illustrated in the previous chapters; the model of the memristor element, $N_0$, was proposed and implemented in Chap. 2; the time-delay parts, $N_1$, are implemented in Chap. 3 from cascade of low-pass second-order Bessel filters; the nonlinear functions $F(X)$, $N_2$ are realized also in Chap. 3 by using a bipolar transistor differential cells.

The two circuits, transmitter (master) and receiver (slave), are time-delay memristive chaotic circuits. The master and slave circuits are practically identical, with the difference being that the transmitter signal $X_{1m}$ replaces the receiver signal $X_{1s}$ as input to the time-delay block.

And referring to the circuit theory and the circuit diagram shown in Fig. 4.11, one can write the following equations:

The master equation

$$\frac{dX_{1m}(t)}{dt} = \frac{1}{R_{1m}C_{1m}}\left(F_m\left((X_{1m})_{\tau_m}\right) - \frac{R_{3m}}{R_{2m}}h(X_{1m}, X_{2m})\right) \qquad (4.12)$$

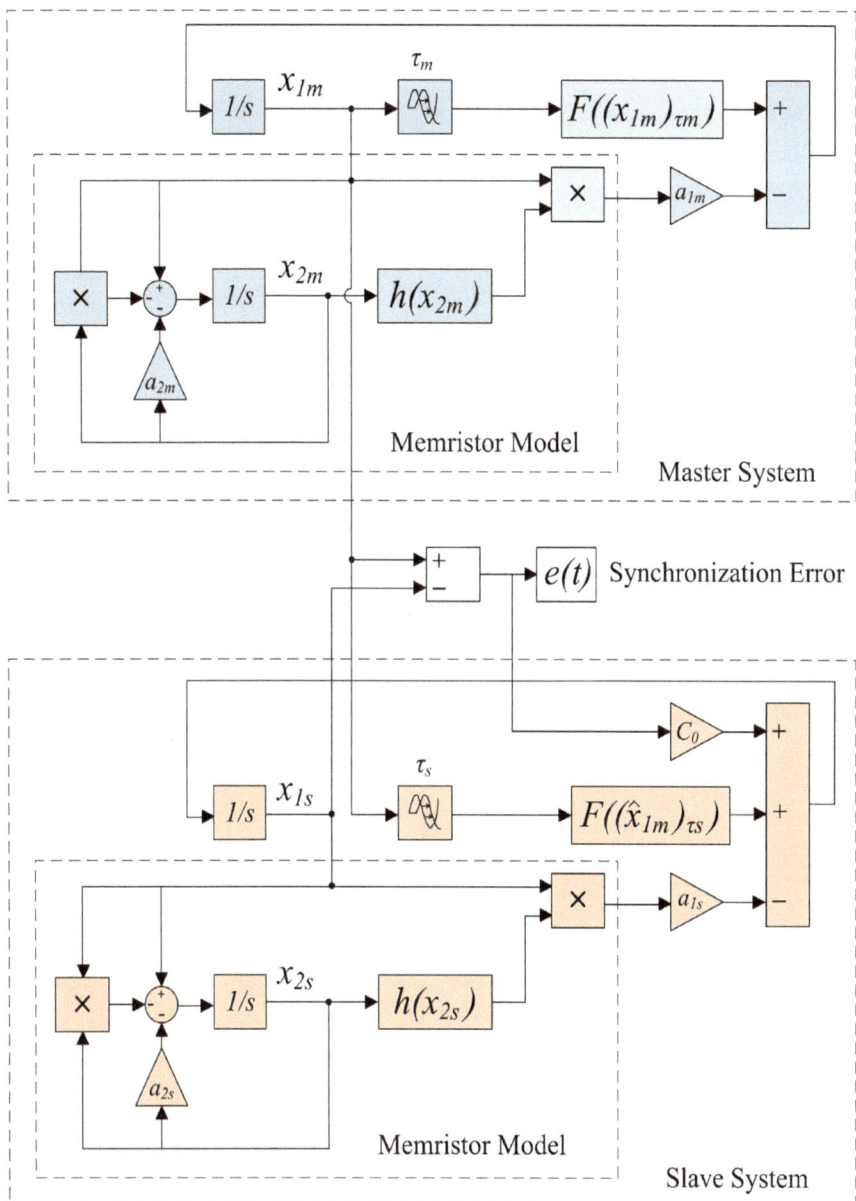

**Fig. 4.7** Simulink model of the feedback control synchronization

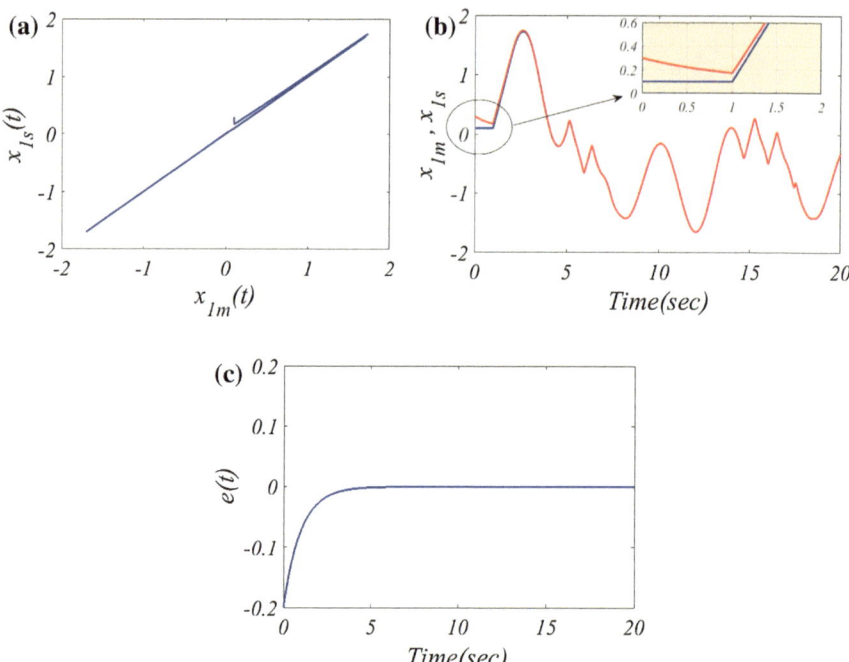

**Fig. 4.8** Feedback control synchronization method between master and slave systems with match parameters, feedback gain $C_0 = 1$ and noiseless channel: **a** $x_{1m}(t)$ versus $x_{1s}(t)$, **b** time series of $x_{1m}(t)$ (blue) and $x_{1s}(t)$ (red), and **c** the error of synchronization

The slave equation

$$\frac{dX_{1s}(t)}{dt} = \frac{1}{R_{1s}C_{1s}}\left(F_s\left(\left(\hat{X}_{1m}\right)_{\tau_s}\right) - \frac{R_{3s}}{R_{2s}}h(X_{1s}, X_{2s})\right) \qquad (4.13)$$

The simulation results have two branches:

- The master and salve circuits have matched elements, the PC synchronization is achieved as shown in Fig. 4.12a, and the synchronization error time series of master and slave circuits Fig. 4.12b is relatively small as shown in Fig. 4.12c. Parameter values: $R_{1m,s} = 10$ k, $R_{2m,s} = R_{3m,s} = 2$ k, $R_{4m,s} = 3$ k, $R_{5m,s} = 1.25$ k, $R_{ich} = 1$ k$(i = 1, 2, 3, 4, 5)$, $C_{1m,s} = 100$ nF.
- The master and salve circuits have mismatched parameters; the PC synchronization method is failed as illustrated in phase portrait and time series of Fig. 4.13a, b and can be observed that the synchronization error is large as shown in Fig. 4.13b which is large. Parameter values: $R_{1m,s} = 10$ k, $R_{2m,s} = R_{3m,s} = 2$ k, $R_{4m} = 3$ k, $R_{4s} = 4$ k, $R_{5m} = 1.25$ k, $R_{5s} = 1.4$ k, $R_{ich} = 1$ k$(i = 1, 2, 3, 4, 5)$, $C_{1m,s} = 100$ nF.

The PC synchronization method has been tested by adding noise $N(t)$ signal to transmitted signal $(S = 1)$. The synchronization is lost at SNR = 5 dB as shown in

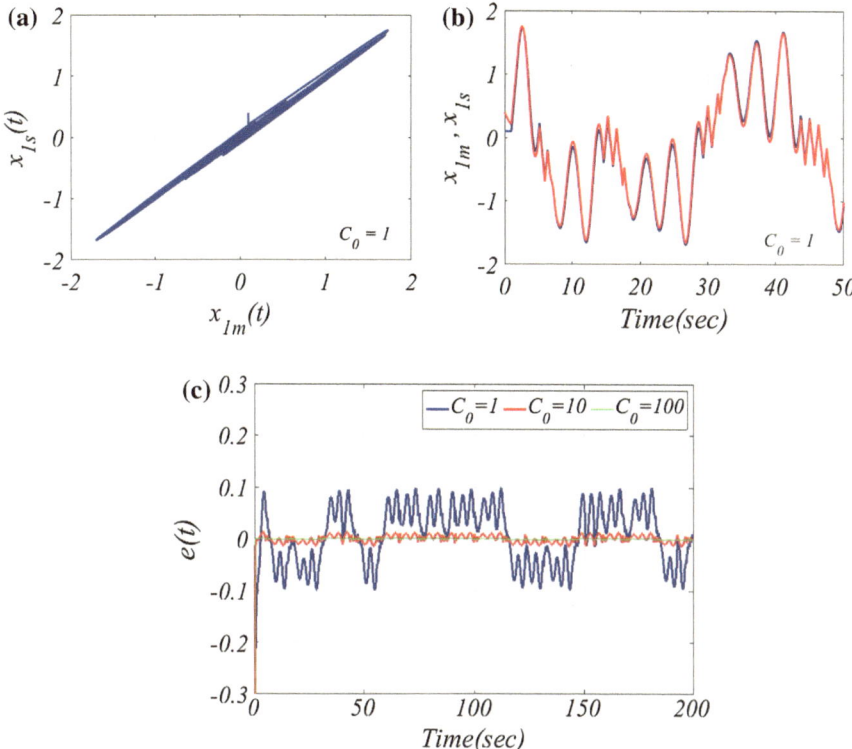

**Fig. 4.9** Feedback control synchronization method between master and slave systems with mismatch parameters and noiseless channel: **a** $x_{1m}(t)$ versus $x_{1s}(t)$, **b** time series of $x_{1m}(t)$ (blue) and $x_{1s}(t)$ (red), and **c** the error of synchronization

Fig. 4.14. By increasing the signal-to-noise ratio to SNR = 10 dB and SNR = 20 dB, the synchronization error decreased as shown in Figs. 4.15 and 4.16, respectively.

### 4.2.2 Feedback Control Synchronization Electronic Circuit

This method of synchronization differs from PC synchronization by adding controller to the receiver circuit. Figure 4.17 shows the circuit diagram. This circuit diagram includes the same parts of PC synchronization circuit with controller. The controller is the electronic realization of Eq. (4.11). The circuital equation is expressed as follow:

$$E(t) = \frac{R_{3c}}{R_{2c} + R_{3c}} \left(1 + \frac{R_{4c}}{R_{1c}}\right) X_{1s} - \frac{R_{4c}}{R_{1c}} X_{1m} \qquad (4.14)$$

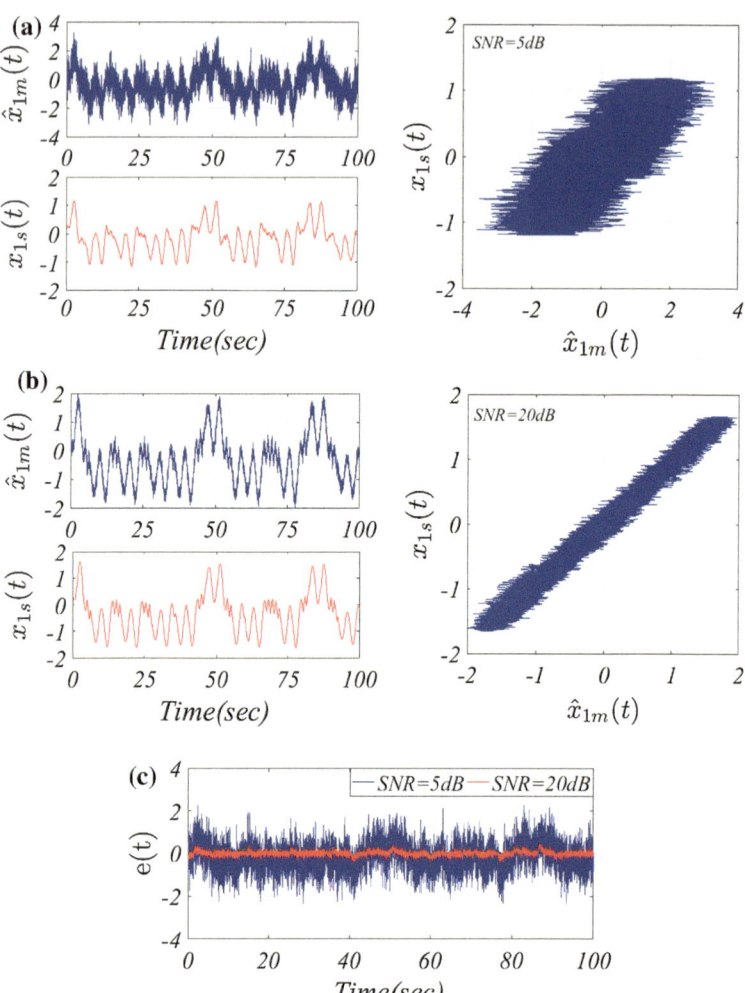

**Fig. 4.10** Time series of master–slave signal and the phase portrait which describes the $\hat{x}_{1m}$ versus $x_{1s}$ of the memristive time-delay paired systems in a feedback control synchronization with match parameters, feedback gain $C_0 = 1$ and noisy channel. **a** SNR = 5 dB, **b** SNR = 20 dB, and **c** synchronization errors

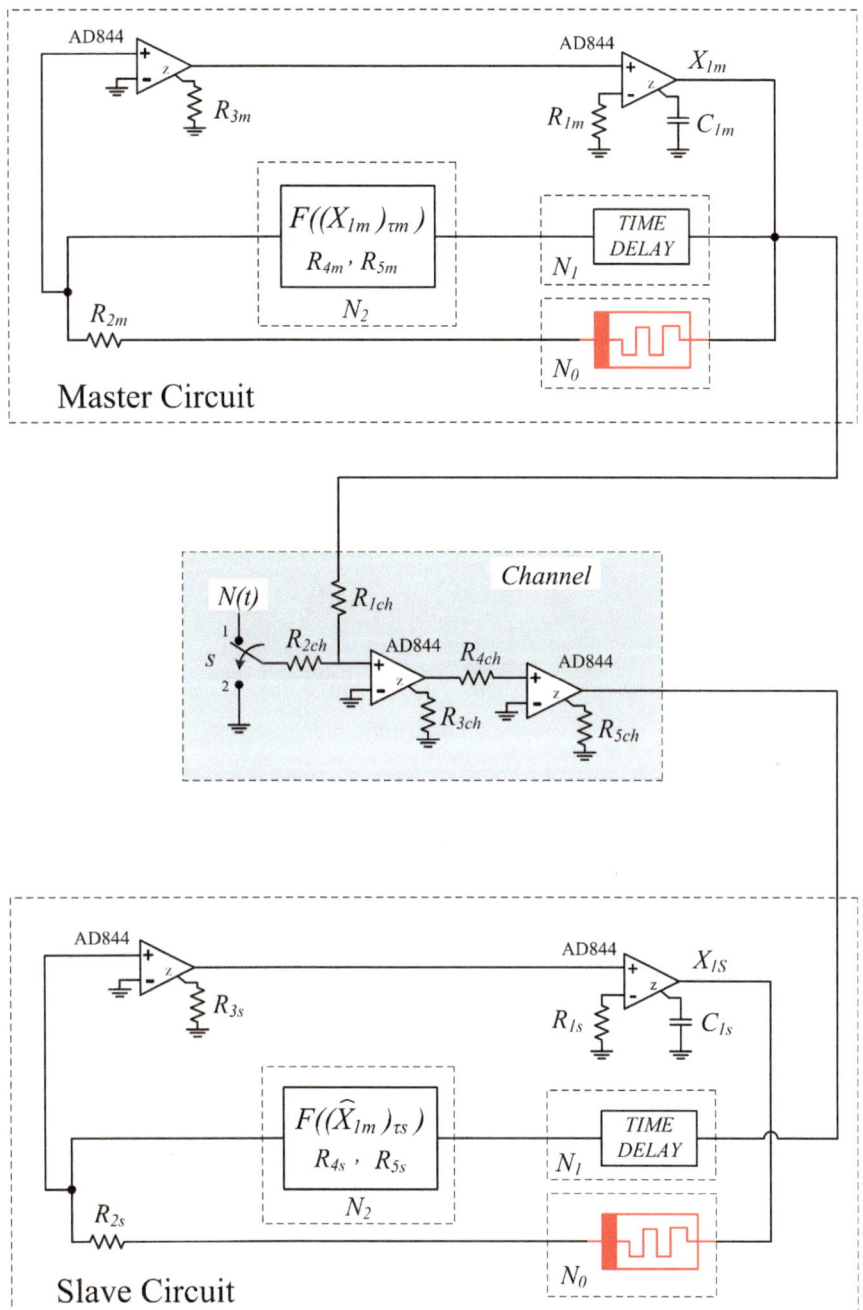

**Fig. 4.11** Scheme of PC synchronization circuit

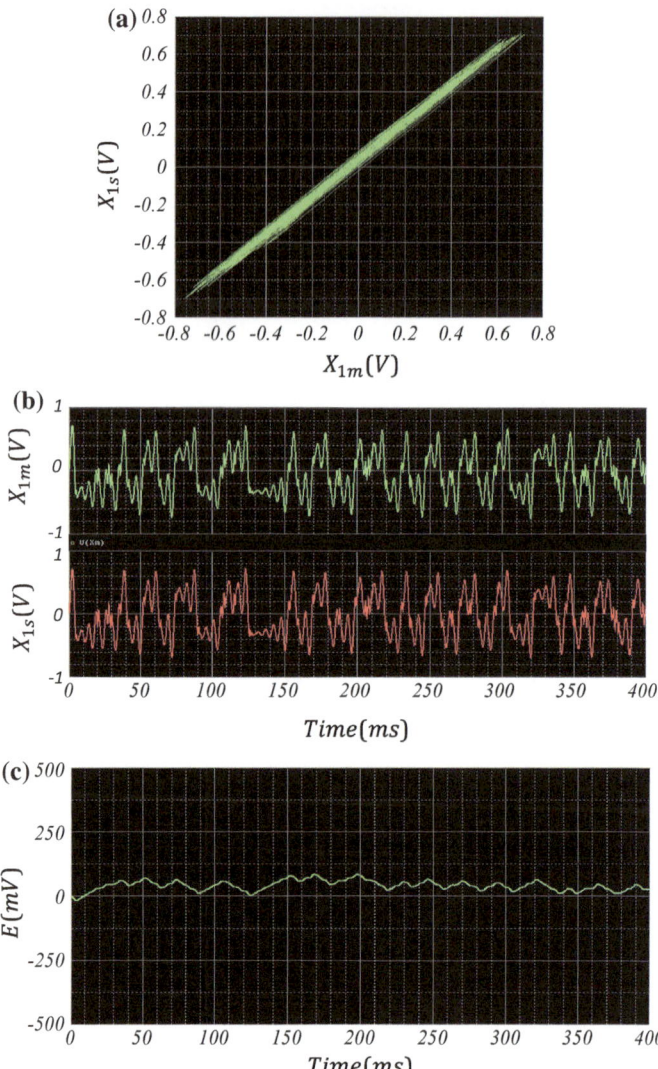

**Fig. 4.12** PSpice results of PC synchronization with match components values and noiseless channel: **a** $X_{1m}(t)$ versus $X_{1s}(t)$, **b** time series of $X_{1m}(t)$ and $X_{1s}(t)$, and **c** the error of synchronization

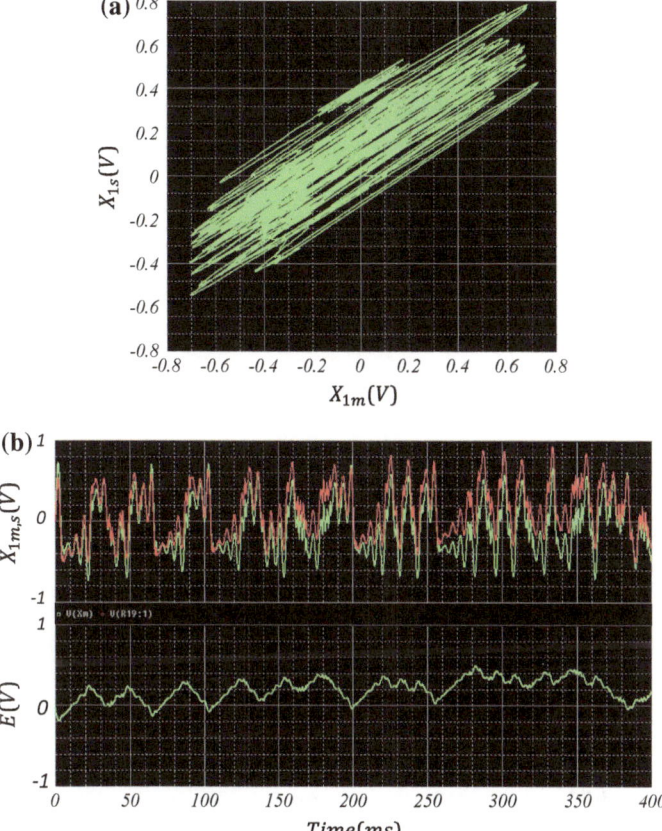

**Fig. 4.13** PSpice results of PC synchronization with mismatch components values and noiseless channel: **a** $X_{1m}(t)$ versus $X_{1s}(t)$, **b** time series of $X_{1m}(t)$ (green) and $X_{1s}(t)$ (red), and the error of synchronization

And referring to the circuit theory and the circuit diagram shown in Fig. 4.17, one can write finally the following equation:

The master equation is:

$$\frac{dX_{1m}(t)}{dt} = \frac{1}{R_{1m}C_{1m}}\left(F_m\big((X_{1m})_{\tau_m}\big) - \frac{R_{3m}}{R_{2m}}h(X_{1m}, X_{2m})\right) \tag{4.15}$$

The slave equation is:

$$\frac{dX_{1s}(t)}{dt} = \frac{1}{R_{1s}C_{1s}}\left(F_s\big((\hat{X}_{1m})_{\tau_s}\big) - \frac{R_{3s}}{R_{2s}}h(X_{1s}, X_{2s}) - \frac{R_{3s}}{R_{5c}}E\right) \tag{4.16}$$

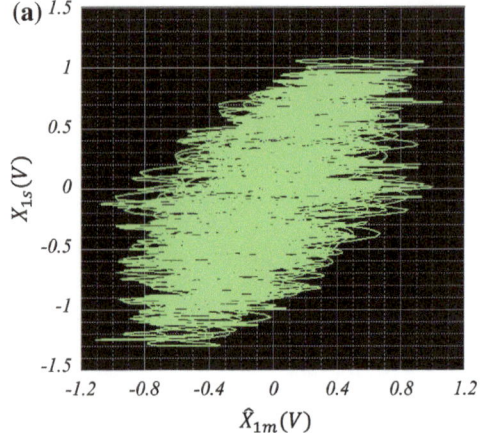

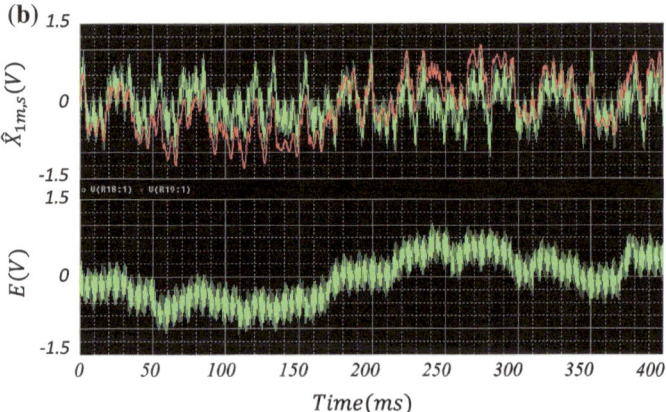

**Fig. 4.14** PSpice results of PC synchronization with matched components values and noisy channel (SNR = 5 dB) **a** $X_{1m}(t)$ versus $X_{1s}(t)$, **b** time series of $X_{1m}(t)$ and $X_{1s}(t)$, and the error of synchronization

The PSpice simulation results have two branches:

- The master–salve circuit has matched elements, the feedback control synchronization is achieved as shown in Fig. 4.18, and the synchronization error between time series of master and slave circuits Fig. 4.18b is relatively small as shown in Fig. 4.18c. Parameter values: $R_{1m,s} = 10$ k, $R_{2m,s} = R_{3m,s} = 2$ k, $R_{4m,s} = 3$ k, $R_{5m,s} = 1.25$ k, $R_{ich} = R_{1c} = R_{2c} = R_{3c} = R_{4c} = 1$ k $(i = 1, 2, 3, 4, 5)$, $R_{5c} = 2$, $0.2, 0.02$ k, $C_{1m,s} = 100$ nF.
- The master–salve circuit has mismatched elements; the feedback control synchronization method will achieve the synchronization between the transmitted signal $\hat{X}_{1m}$ and the receiver signal $X_{1s}$ as illustrated in phase portrait and time series of Fig. 4.19. Figure 4.19c illustrates the synchronization error; it is oscil-

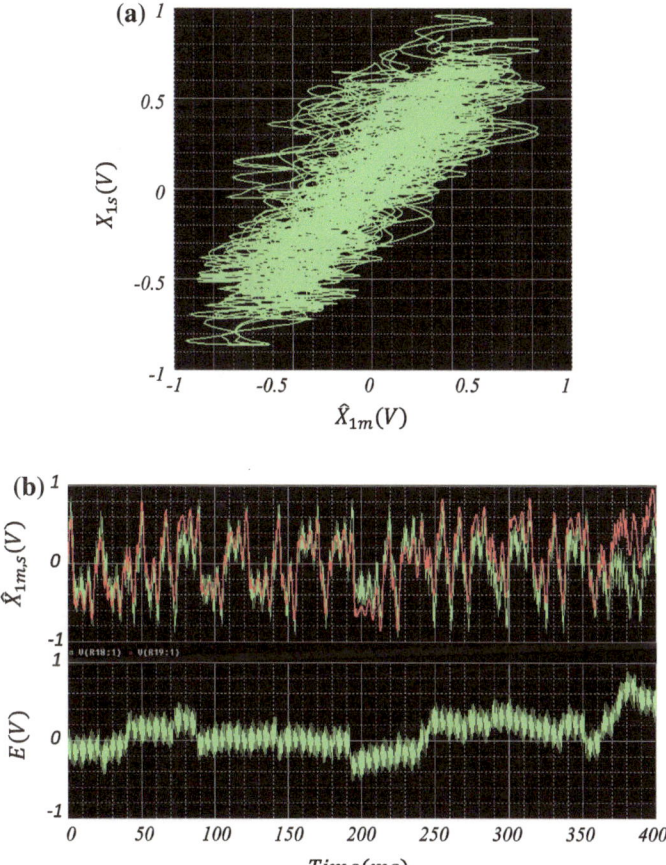

**Fig. 4.15** PSpice results of PC synchronization with matched components values and noisy channel (SNR $= 10$ dB) **a** $X_{1m}(t)$ versus $X_{1s}(t)$, **b** time series of $X_{1m}(t)$ and $X_{1s}(t)$, and the error of synchronization

lating about zero. The synchronization error depends on the coupling strength factor $\left(C_0 = \frac{R_{3s}}{R_{5c}}\right)$. By selecting $C_0 = 1 (R_{3s} = 2$ k, $R_{5c} = 2$ k), the synchronization error observed is large than the case of coupling strength factor $C_0 = 10$ ($R_{3s} = 2$ k, $R_{5c} = 0.2$ k) and the error of the previous two cases ($C_0 = 1, C_0 = 10$) is large than the case of $C_0 = 100$ ($R_{3s} = 2$ k, $R_{5c} = 0.02$ k); i.e., increasing the coupling strength factor $C_0$ will lead to decreasing the synchronization error as shown in Figs. 4.19, 4.20 and 4.21. The values of the components are selected as: $R_{1m,s} = 10$ k, $R_{2m,s} = R_{3m,s} = 2$ k, $R_{4m} = 3$ k, $R_{4s} = 4$ k, $R_{5m} = 1.25$ k, $R_{5s} = 1.4$ k, $R_{ich} = R_{1c} = R_{2c} = R_{3c} = R_{4c} = 1$ k ($i = 1, 2, 3, 4, 5$), $R_{5c} = 2, 0.2, 0.02$ k, $C_{1m,s} = 100$ nF.

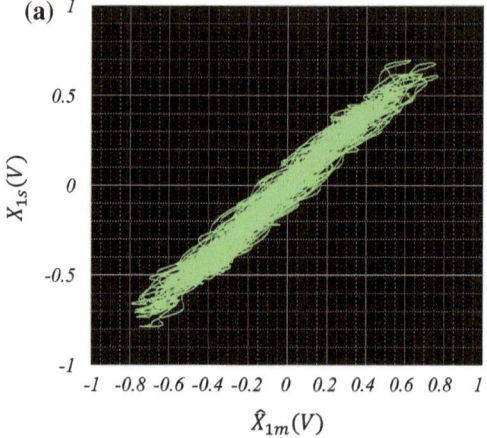

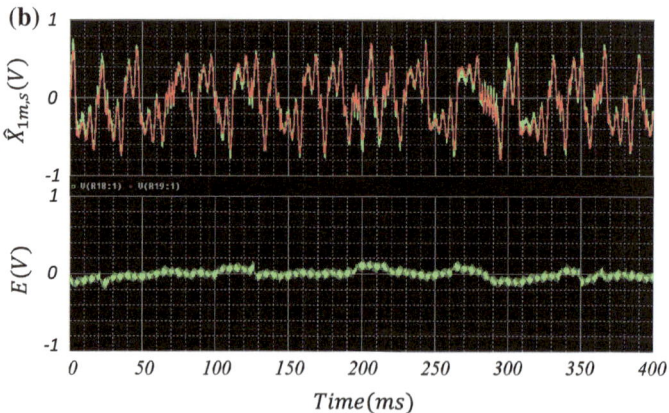

**Fig. 4.16** PSpice results of PC synchronization with matched components values and noisy channel (SNR $= 20$ dB) **a** $X_{1m}(t)$ versus $X_{1s}(t)$, **b** time series of $X_{1m}(t)$ and $X_{1s}(t)$, and the error of synchronization

The feedback control synchronization method has been tested by adding noise $N(t)$ signal to transmitted signal ($S = 1$). The synchronization error observed at SNR $= 5$ dB is high as shown in Fig. 4.22. By increasing the signal-to-noise ratio to SNR $= 10$ dB and SNR $= 20$ dB, the synchronization error decreased as shown in Fig. 4.23 and Fig. 4.24, respectively. Table 4.1 shows performance comparison of PC and feedback control methods, which shows that the performance of feedback control is better than PC (Fig. 4.25).

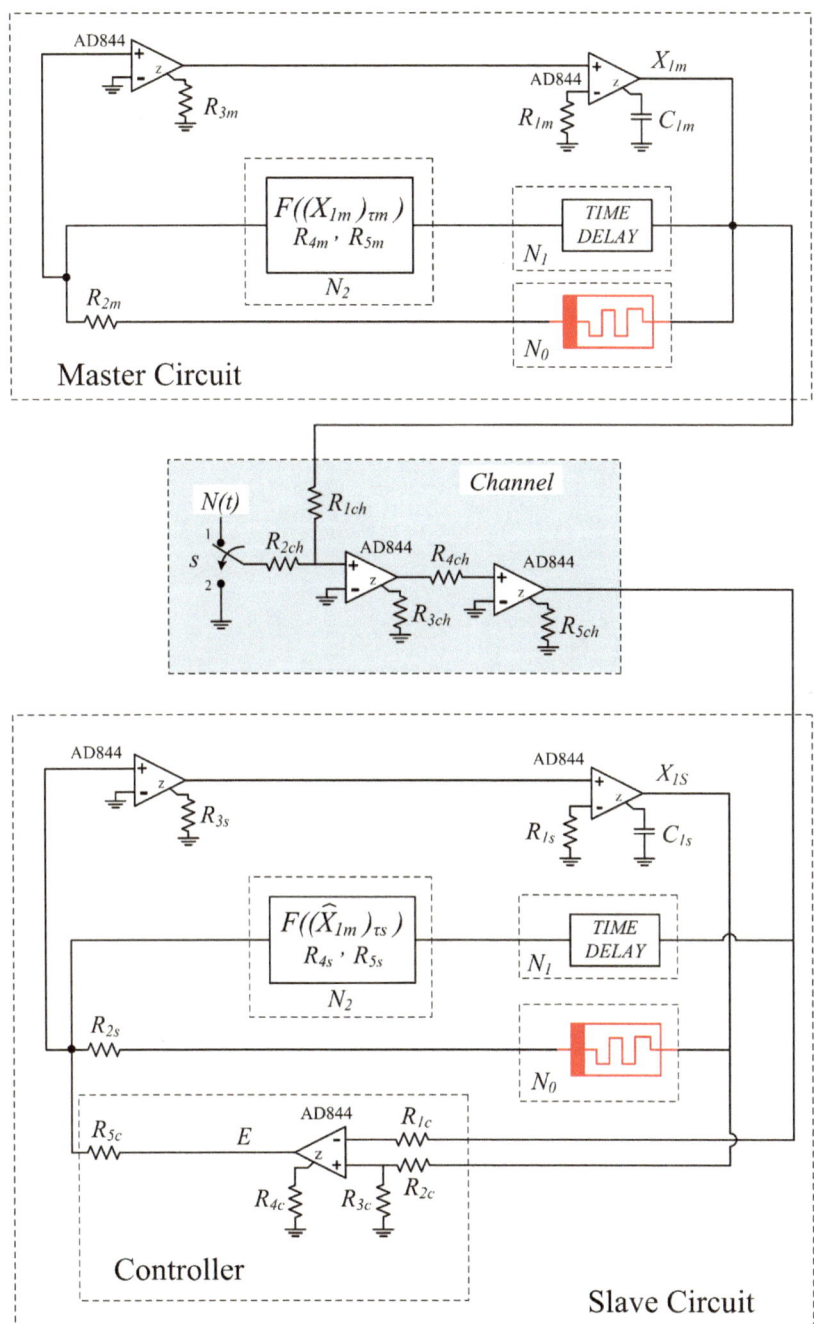

**Fig. 4.17** Scheme of feedback control synchronization circuit

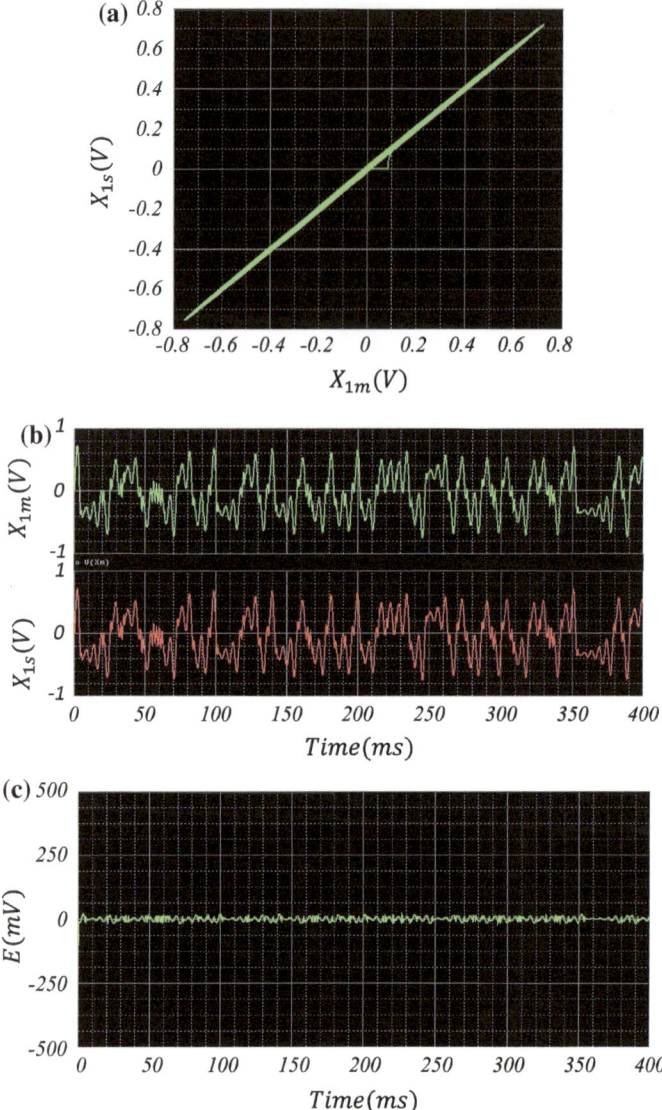

**Fig. 4.18** PSpice results of feedback control synchronization with matched components values, feedback gain $C_0 = 1$ and noiseless channel: **a** $X_{1m}(t)$ versus $X_{1s}(t)$, **b** time series of $X_{1m}(t)$ and $X_{1s}(t)$, **c** the error of synchronization

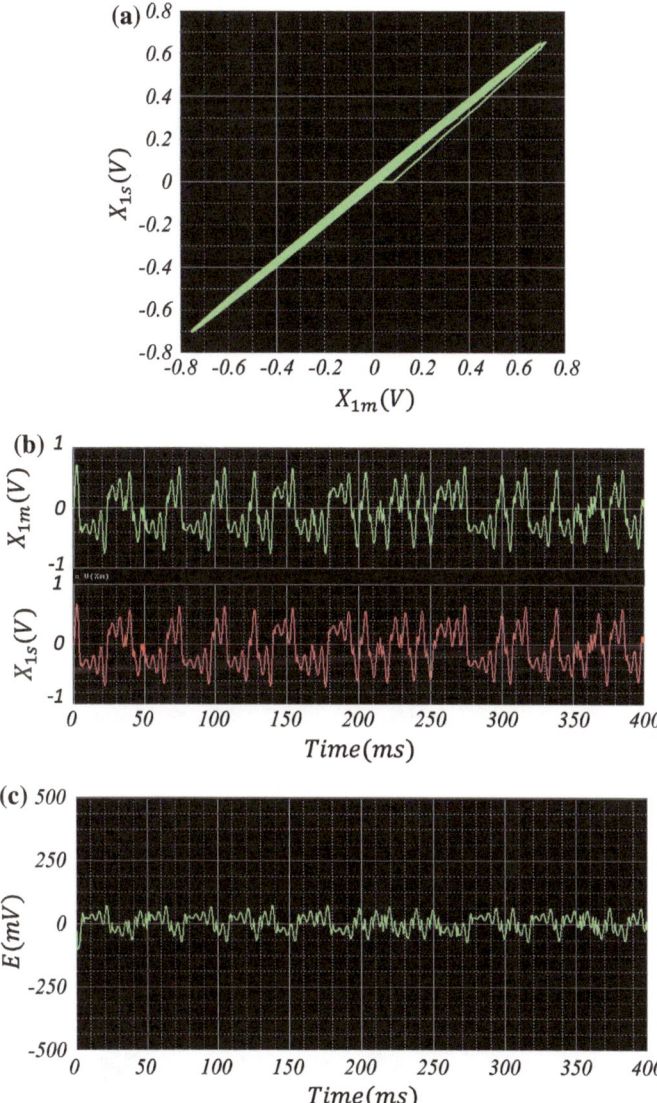

**Fig. 4.19** PSpice results of feedback control synchronization with mismatched components values, feedback gain $C_0 = 1$ and noiseless channel: **a** $X_{1m}(t)$ versus $X_{1s}(t)$, **b** time series of $X_{1m}(t)$ and $X_{1s}(t)$, and **c** the error of synchronization

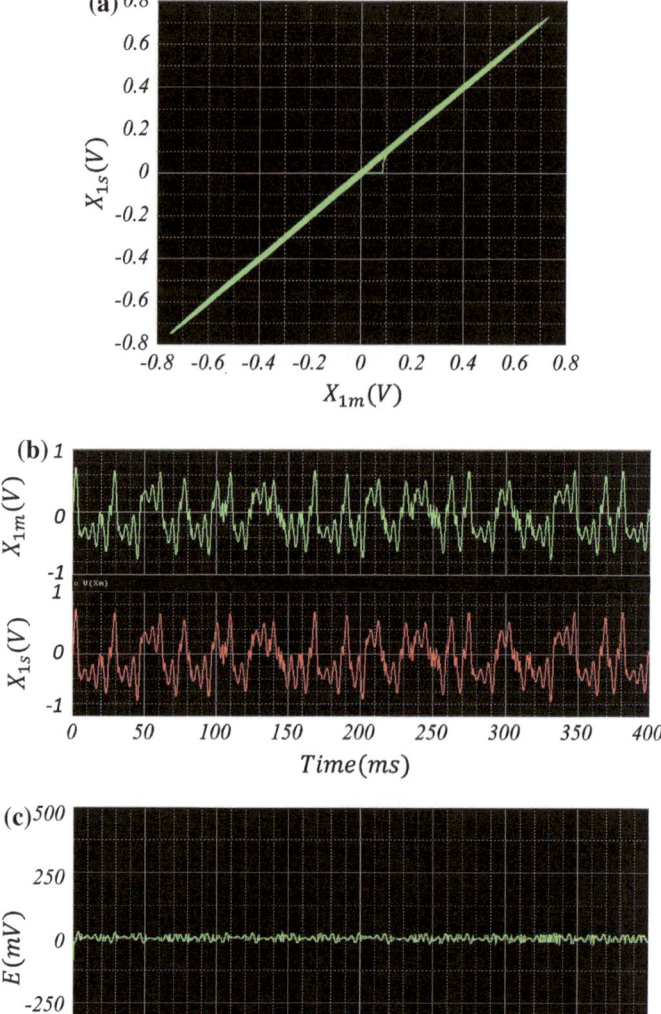

**Fig. 4.20** PSpice results of feedback control synchronization with mismatched components values, feedback gain $C_0 = 10$ and noiseless channel: **a** $X_{1m}(t)$ versus $X_{1s}(t)$, **b** time series of $X_{1m}(t)$ and $X_{1s}(t)$, and **c** the error of synchronization

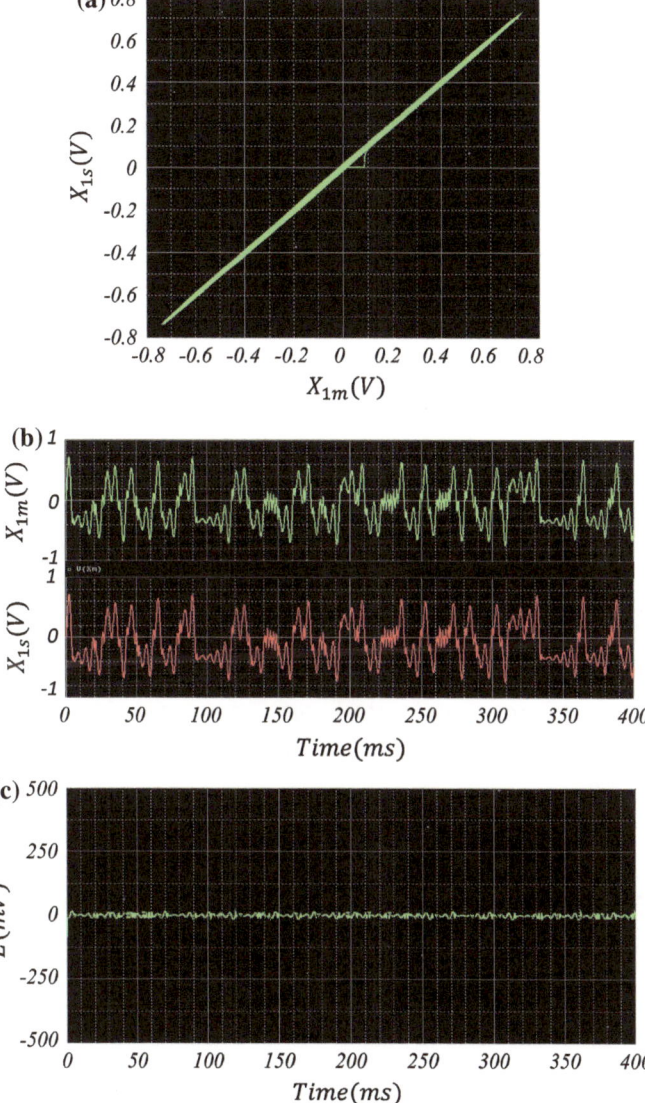

**Fig. 4.21** PSpice results of feedback control synchronization with mismatched components values, feedback gain $C_0 = 100$ and noiseless channel: **a** $X_{1m}(t)$ versus $X_{1s}(t)$, **b** time series of $X_{1m}(t)$ and $X_{1s}(t)$, and **c** the error of synchronization

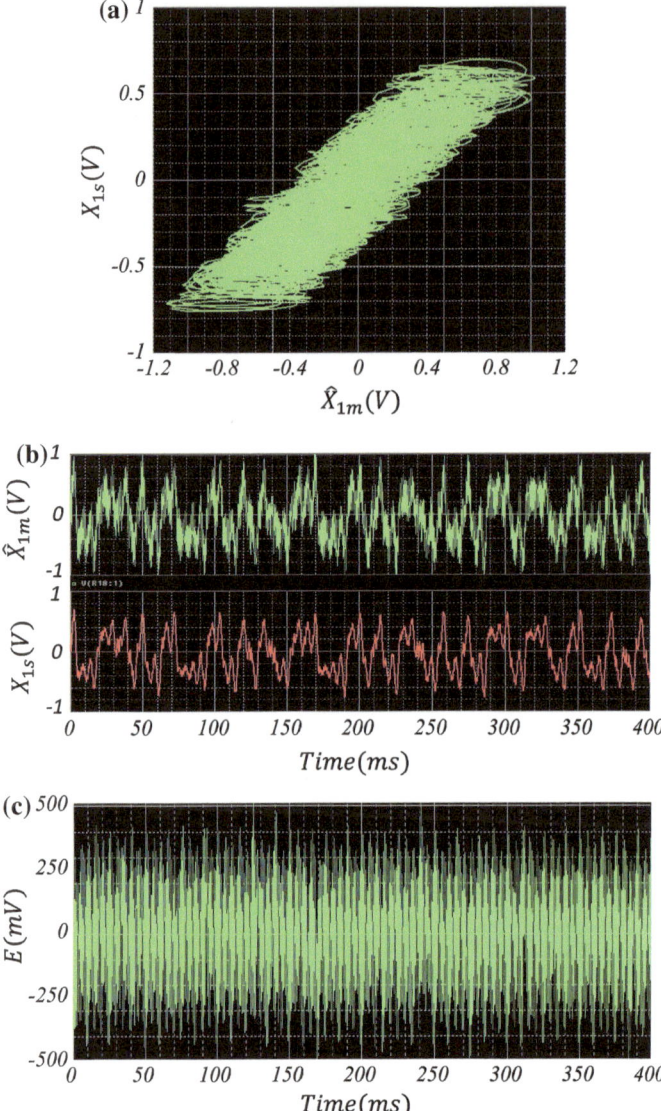

**Fig. 4.22** PSpice results of feedback control synchronization with matched components values, feedback gain $C_0 = 1$ and noisy channel (SNR $= 5$ dB) **a** $\hat{X}_{1m}(t)$ versus $X_{1s}(t)$, **b** Time series of $\hat{X}_{1m}(t)$ and $X_{1s}(t)$, and **c** the error of synchronization

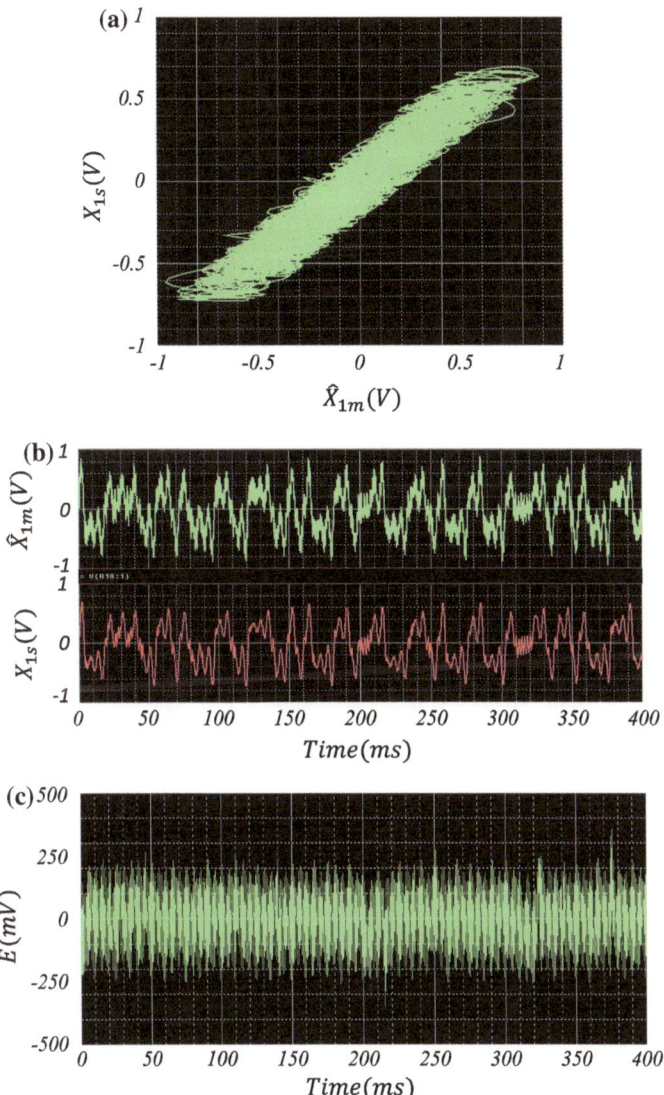

**Fig. 4.23** PSpice results of feedback control synchronization with matched components values, feedback gain $C_0 = 1$ and noisy channel (SNR $= 10$ dB) **a** $\hat{X}_{1m}(t)$ versus $X_{1s}(t)$, **b** time series of $\hat{X}_{1m}(t)$ and $X_{1s}(t)$, and **c** the error of synchronization

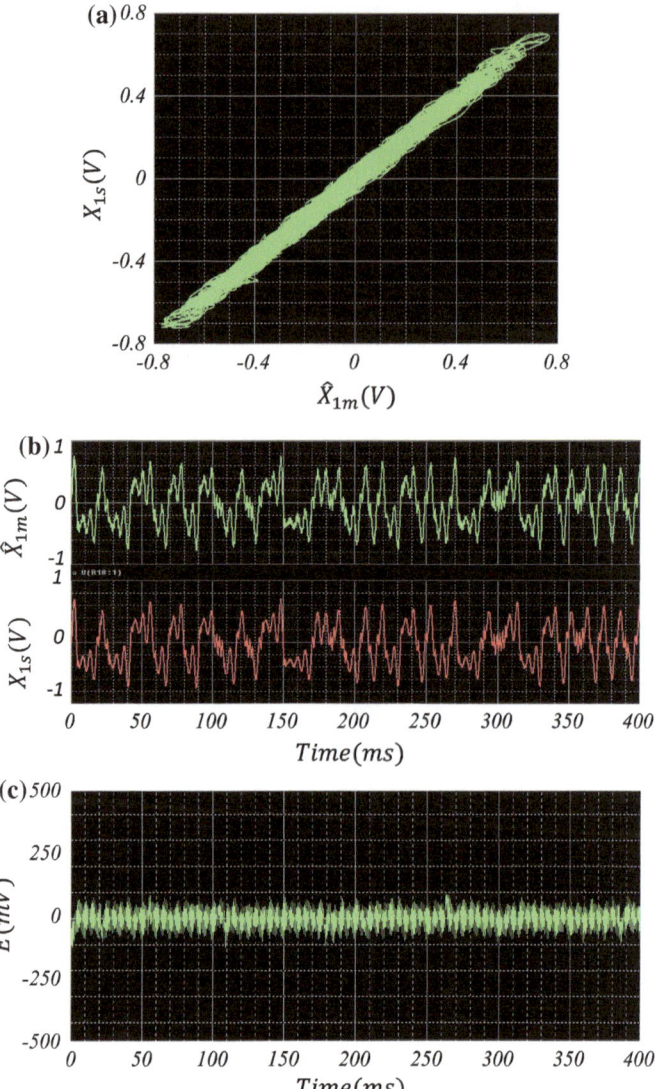

**Fig. 4.24** PSpice results of feedback control synchronization with matched components values, feedback gain $C_0 = 1$ and noisy channel (SNR = 20 dB) **a** $\hat{X}_{1m}(t)$ versus $X_{1s}(t)$, **b** time series of $\hat{X}_{1m}(t)$ and $X_{1s}(t)$, and **c** the error of synchronization

**Table 4.1** Synchronization error between the two methods

| SNR (dB) | PC synchronization | Feedback control synchronization | | |
|---|---|---|---|---|
| | Maximum error (V) | Maximum error (V) | | |
| | | $c_0 = 1$ | $c_0 = 10$ | $c_0 = 100$ |
| 5 | 1 | 0.5 | 0.42 | 0.12 |
| 10 | 0.9 | 0.35 | 0.27 | 0.09 |
| 20 | 0.2 | 0.16 | 0.08 | 0.038 |

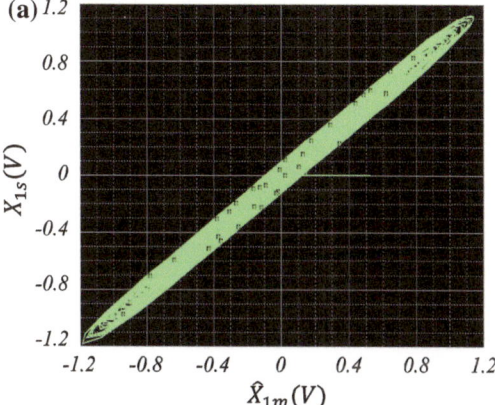

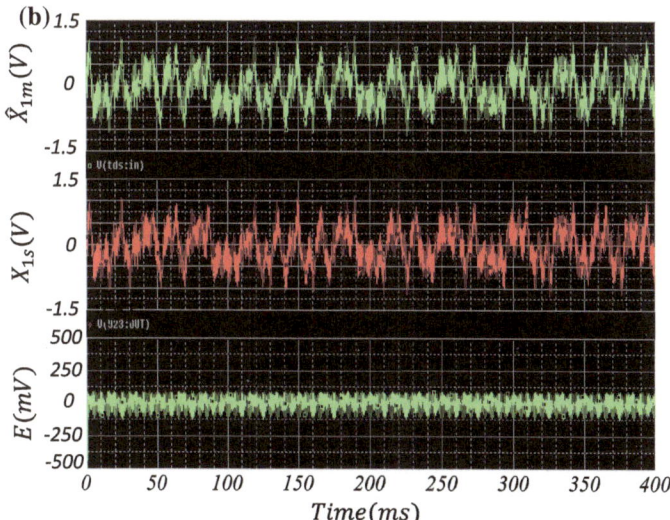

**Fig. 4.25** PSpice results of feedback control synchronization with mismatched components values, feedback gain $C_0 = 100$ and noisy channel (SNR = 20 dB) **a** $\hat{X}_{1m}(t)$ versus $X_{1s}(t)$, **b** time series of $\hat{X}_{1m}(t)$, $X_{1s}(t)$, and the error of synchronization

# References

1. T.L. Carroll, L.M. Pecora, Synchronization in chaotic systems. Phys. Rev. Lett. **64**(8), 215–248 (1990)
2. A. Pikovsky, M. Rosenblum, J. Kurths, Synchronization: A universal concept in nonlinear sciences, Cambridge University press, ISBN 978-0-511-07595-7 (2003)
3. K. Ding, C. Volos, X. Xu, B. Du, Master-Slave synchronization of 4D hyperchaotic rabinovich systems, vol. 2018 (Wiley Hindawi, 2018)
4. S. Vaidyanathan, Adaptive control and synchronization of Rossler prototype-4 system. Int. J. Adv. Inf. Technol. **1**(5), 11 (2011)
5. W. He, F. Qian, Q.-L. Han, J. Cao, Lag quasi-synchronization of coupled delayed systems with parameter mismatch. IEEE Trans. Circuits Syst. I Regul. Pap. **58**(6), 1345–1357 (2011)
6. C. Liu, C. Li, C. Li, Quasi-synchronization of delayed chaotic systems with parameters mismatch and stochastic perturbation. Commun. Nonlinear Sci. Numer. Simul. **16**(10), 4108–4119 (2011)

# Chapter 5
# Cryptography Based on Memristive Electronic Circuits

Cryptography is the science that uses the calculation and math behind the procedures to encrypt and decrypt data. The process that is used to convert the message (text, picture, sound, and video) into cipher message is known as encryption. The encryption algorithm is used to encrypt a message at the transmitter, while the decryption algorithm is applied to decrypt the received encrypted message. In a cryptosystem, the synchronization of the transmitter and receiver sides has to be secured. One can use a key during encryption and decryption process. According to the key, the cryptography is classified into two kinds:

- Symmetric key: the transmitter and receiver use the same key.
- Asymmetric key: different keys are used for encryption and decryption.

In chaotic cryptographic technique, the symmetric key has been used [1].

This chapter is organized as follows: The cryptography technique is proposed, which is used $n$-shift cipher algorithm for encrypting and decrypting messages. The simulation results of the crypto-circuits have been observed. The design and simulation have been confirmed by using PSpice.

## 5.1 The Proposed Cryptography Technique

This technique is a modified method that had been proposed in [2]. Figure 5.1 shows the block diagram of the proposed method. The transmitter (master) consists of two memristive chaotic circuits and an encryption function $e(.)$, and the receiver (slave) also has two memristive chaotic circuits and a decryption function $d(.)$. The memristive chaotic circuit $m_1$ generates two output signals, $y_1(t)$ and $y_2(t)$. The signal $y_1(t)$ is used for driving the second memristive chaotic circuit $m_2$, while the output $y_2(t)$ is used for masking purposes. The keystream $k(t)$ is one of the state variables of the memristive chaotic circuit $m_2$ which is used for encrypting the message $m(t)$

© The Author(s), under exclusive license to Springer Nature Switzerland AG 2019
F. Rahma and S. Muneam, *Memristive Nonlinear Electronic Circuits*,
SpringerBriefs in Nonlinear Circuits, https://doi.org/10.1007/978-3-030-11921-8_5

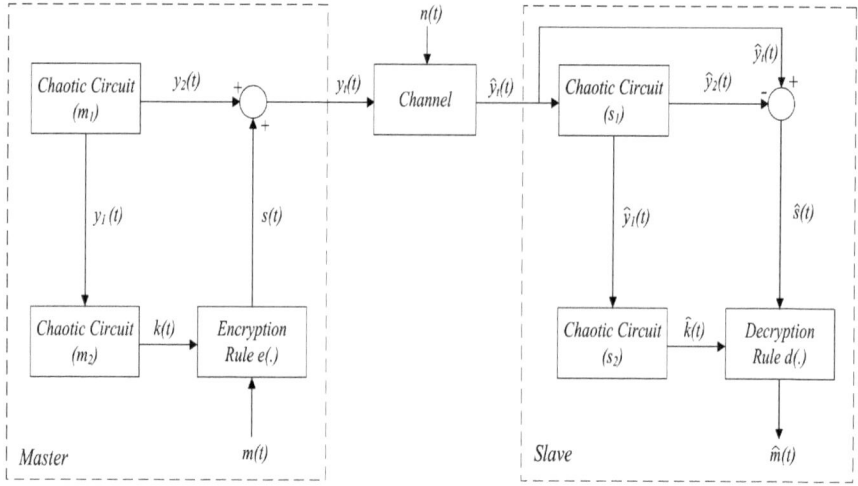

**Fig. 5.1** Block diagram of communication system based on cryptography technique

by using the rule $e(.)$. The encrypted signal $s(t)$ is masked using $y_2(t)$ yielding the transmitted signal $y_t(t)$. In the real channel, the transmitted signal $y_t(t)$ is subjected to the noise $n(t)$.

Two synchronization techniques have been used in the common scheme:

(i)  Direct synchronization exists between two memristive chaotic circuits $m_1$ and $s_1$ via feedback control method. So, the chaotic circuit $s_1$ can estimate $\hat{y}_1(t)$ and $\hat{y}_2(t)$ of the signals $y_1(t)$ and $y_2(t)$, respectively. The estimated signal $\hat{s}(t)$ can be generated by unmasking process.

(ii)  Indirect synchronization exists between two memristive chaotic circuits $m_2$ and $s_2$. According to the indirect synchronization, the keystream $\hat{k}(t)$ has been estimated. Consequently, the decryption rule $d(.)$ is applied to recover the original message $m(t)$.

## 5.1.1  The Cipher Key Algorithm

$n$-shift cipher algorithm has been used for encryption and decryption purposes. The encryption algorithm is given as follows [2]:

$$e(m(t), k(t)) = \underbrace{f_1(\ldots f_1(f_1(m(t), k(t)), k(t)), \ldots, k(t))}_{n} = s(t) \qquad (5.1)$$

**Fig. 5.2** Nonlinear function used in continuous $n$-shift cipher

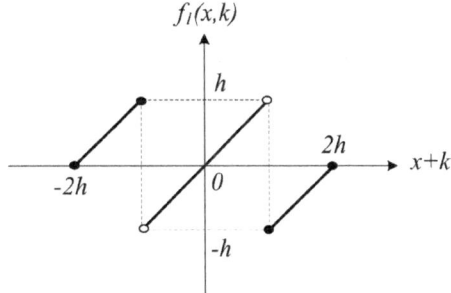

where $f_1$ is a nonlinear function. The nonlinear function is shown in Fig. 5.2 and given by:

$$f_1(x, k) = \begin{cases} (x+k) + 2h, & -2h \le (x+k) \le -h \\ (x+k), & -h < (x+k) < h \\ (x+k) - 2h, & h \le (x+k) \le 2h \end{cases} \quad (5.2)$$

where $h$ is the encryption parameter. The message $m(t)$ and key $k(t)$ should lie within the interval $(-h, h)$.

At the receiver side, the decryption function $d(.)$ has been used, which is the same of the encryption rule. The decryption rule is given as:

$$\hat{m}(t) = d\left(\hat{s}(t), \hat{k}(t)\right)$$
$$= f_1\left(\cdots f_1\left(f_1\left(\hat{s}(t), -\hat{k}(t)\right), -\hat{k}(t)\right), \cdots, -\hat{k}(t)\right) \quad (5.3)$$

where $\hat{k}(t)$ is the estimated keystream.

According to the communication scheme clarified by Fig. 5.1, the dynamical model (3.16) is used as transmitter and receiver systems $m_1, s_1$, respectively, and can be described as follows:

$$m_1 : \begin{cases} \dot{x}_{1m}(t) = \sigma x_{2m} \\ \dot{x}_{2m}(t) = \beta h(x_{1m}, x_{4m}) + \alpha x_{3m} - \gamma x_{2m} \\ \dot{x}_{3m}(t) = 1 - \alpha h(x_{3m}, x_{5m}) - x_{1m} \\ \dot{x}_{4m}(t) = x_{1m} - x_{1m} x_{4m} - \delta x_{4m} \\ \dot{x}_{5m}(t) = x_{3m} - x_{3m} x_{5m} - \delta x_{5m} \\ y_1(t) = x_{1m} \\ y_2(t) = x_{2m} \\ y_t(t) = y_2 + s(t) \end{cases} \quad (5.4)$$

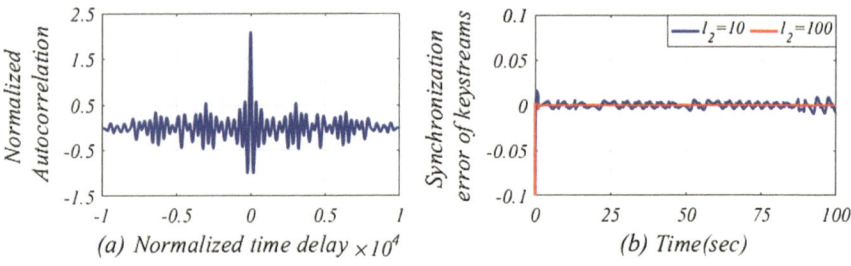

**Fig. 5.3** **a** Correlation of keystream with feedback gain over a noisy channel ($l_2 = 10$) and **b** synchronization error in the estimation of the keystream

$$
s_1 : \begin{cases}
\dot{x}_{1s}(t) = \sigma x_{2s} + c_0(y_1 + n(t) - x_{1s}) \\
\dot{x}_{2s}(t) = \beta h(x_{1s}, x_{4s}) + \alpha x_{3s} - \gamma x_{2s} + c_0(y_t + n(t) - x_{2s}) \\
\dot{x}_{3s}(t) = 1 - \alpha h(x_{3s}, x_{5s}) - x_{1s} \\
\dot{x}_{4s}(t) = x_{1s} - x_{1s}x_{4s} - \delta x_{4s} \\
\dot{x}_{5s}(t) = x_{3s} - x_{3s}x_{5s} - \delta x_{5s} \\
\hat{y}_1(t) = x_{1s} \\
\hat{y}_2(t) = x_{2s}
\end{cases}
\tag{5.5}
$$

The dynamical model (3.2) is used for implementing the circuits of $m_2$ and $s_2$, and can be described as follows:

$$
m_2 : \begin{cases}
\dot{z}_{1m}(t) = F\big((y_1(t))_{\tau_m}\big) - a_1 h(z_{1m}, z_{2m}) + l_2(z_{1m} - y_1(t)) \\
\dot{z}_{2m}(t) = z_{1m} - z_{1m}z_{2m} - a_2 z_{2m} \\
k(t) = z_{1m}
\end{cases}
\tag{5.6}
$$

$$
s_2 : \begin{cases}
\dot{z}_{1s} = F\big((\hat{y}_1(t))_{\tau_s}\big) - a_1 h(z_{1s}, z_{2s}) + l_2\big(z_{1m} - \hat{y}_1(t)\big) \\
\dot{z}_{2s} = z_{1s} - z_{1s}z_{2s} - a_2 z_{2s} \\
\hat{k} = z_{1s}
\end{cases}
\tag{5.7}
$$

where $l_2$ and $c_0$ are the feedback gains.

The parameter values of the chaotic systems are mentioned in the previous chapters. The encryption parameter $h$ is selected as 1 and the encryption rule ($n$-shift cipher; $n = 2$) is used. Figure 5.3 illustrates the autocorrelation function of the keystream $k(t)$. The keystream is not similar to itself with any amount of time shift so its autocorrelation function has only a single spike at point of zero-time shift. This means the keystream generated is chaotic in nature, and therefore has limited predictability.

The cryptosystem has been used to cipher two kinds of messages:

i. Encrypt analog signal: The speech from the following phrase:

> "In the name of God, the Merciful
>
> Yaa Siiin ⚙ By the Quran, full of the Wisdom ⚙ Thou art indeed one of
> the apostles ⚙ On a straight Way ⚙ It is a revelation sent down by him,
> the exalted in might, most merciful ⚙ In order that thou mayest admonish a
> people, whose fathers had received no admonition, and who therefore
> remain heedless"

has been considered as the original message $m(t)$. Figures 5.4 and 5.5 illustrate speech waveforms $m(t)$, transmitted signal $y_t(t)$ and recovered message $\hat{m}(t)$ with mismatch parameters, noiseless, and noisy (SNR $= 35$ dB) channel, respectively. The filtering process used a low-pass filter (LPF) for excluding the noise.

ii. Encrypt digital signal: The original $m(t)$ is a pseudo-random (PN) code sequence which generates a binary information message as $2^7 - 1(1011011)$. Figures 5.6 and 5.7 illustrate two cases: noiseless and noisy (SNR $= 35$ dB) channels, respectively. Due to the noisy channel effect, the recovered message is corrupted.

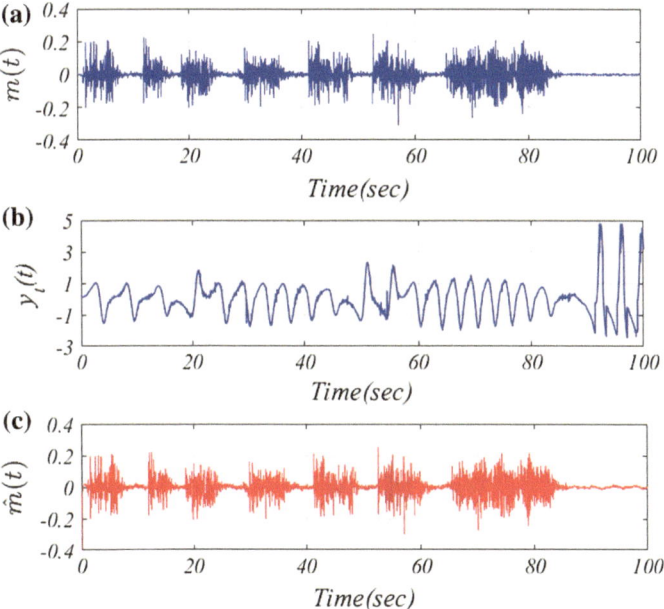

**Fig. 5.4** Secure communication for speech waveform with mismatch parameters, feedback gains $C_0 = 100$, $l_2 = 10$, and noiseless channel **a** original message, **b** transmitted signal, and **c** recovered message

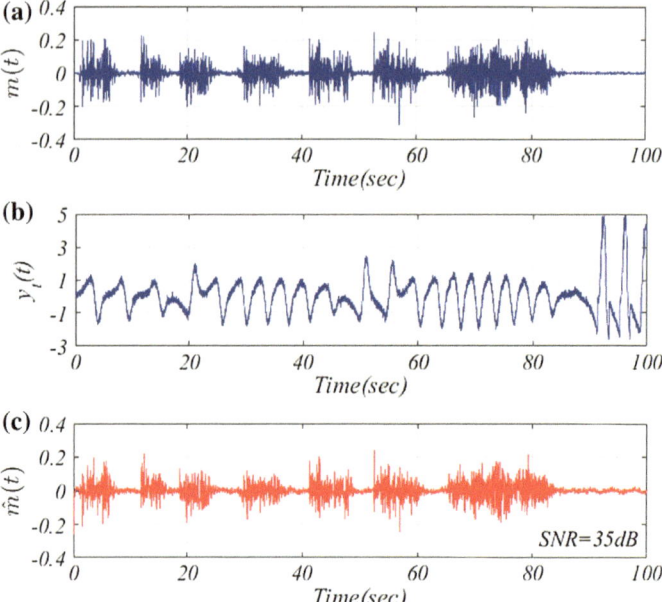

**Fig. 5.5**  Secure communication for speech waveforms with mismatch parameters, feedback gains $C_0 = 100$, $l_2 = 10$, and noisy channel (SNR = 35 dB) **a** original message, **b** transmitted signal, and **c** recovered message after filtering

## 5.2  Circuit Realization

In this section, an electronic circuit designs and implements the secure communication using memristive cryptosystem that is introduced. The circuit diagram shown in Fig. 5.8 consists of three original parts: Part I: four memristive chaotic circuits $(m_1, m_2, s_1, s_2)$. Part II: masking–unmasking circuit. Part III: encryption–decryption circuit. The circuits are designed by using CFOA. In Part I, the circuits are designed and implemented in chapter three. The encryption and decryption blocks are implemented by the circuitry of Fig. 5.9 as given below:

For encryption circuit:

$$\text{sum}_n = -\left(\frac{R_{3n}}{R_{1n}} K(t) + \frac{R_{3n}}{R_{2n}} S_{n-1}(t)\right), \quad n = 1, 2. \tag{5.8}$$

$$S_n(t) = -\frac{|V_{\text{sat}}|}{R_{6n}} \text{sgn}\left[\text{sum}_n - E_j\right] + \frac{\text{sum}_n}{R_{7n}}, \quad j = \pm 1. \tag{5.9}$$

For decryption circuit:

$$\text{dif}_n = \left(\frac{R_{3n}}{R_{2n} + R_{3n}}\left(1 + \frac{R_{8n}}{R_{1n}}\right)\hat{S}_n(t) - \frac{R_{8n}}{R_{1n}}\hat{K}(t)\right), \quad n = 1, 2. \tag{5.10}$$

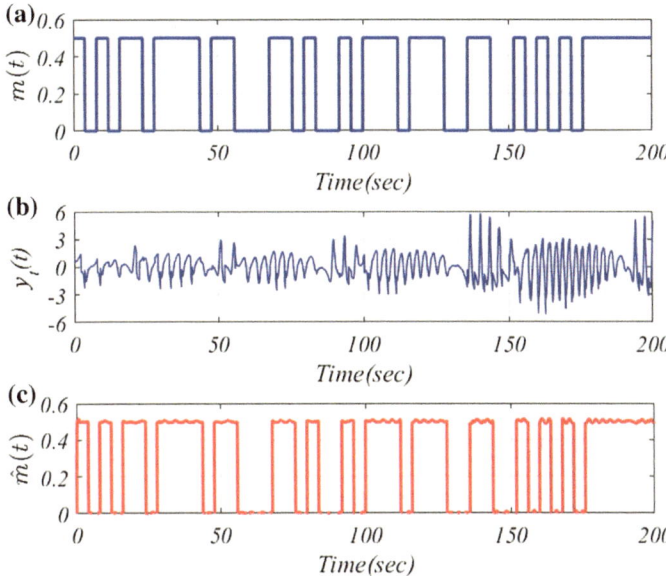

**Fig. 5.6** Secure communication for digital message with mismatch parameters, feedback gains $C_0 = 100$, $l_2 = 10$, and noiseless channel **a** original message, **b** transmitted signal, and **c** recovered message

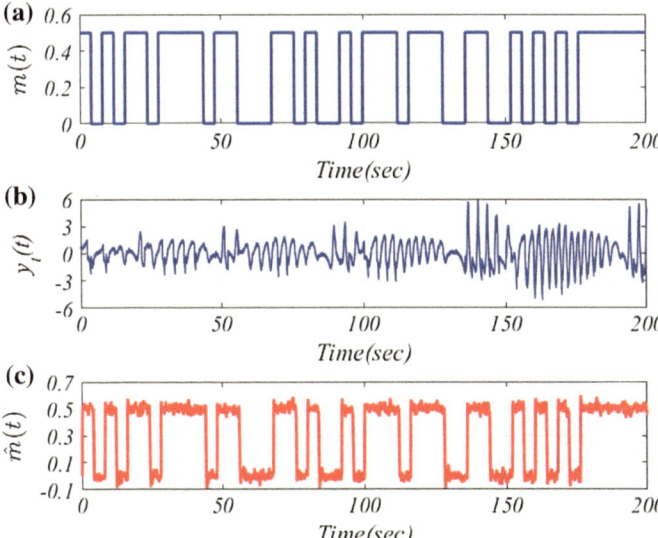

**Fig. 5.7** Secure communication for digital message with mismatch parameters, feedback gains $C_0 = 100$, $l_2 = 10$, and noisy channel (SNR = 35 dB) **a** original message, **b** transmitted signal, and **c** recovered message after filtering

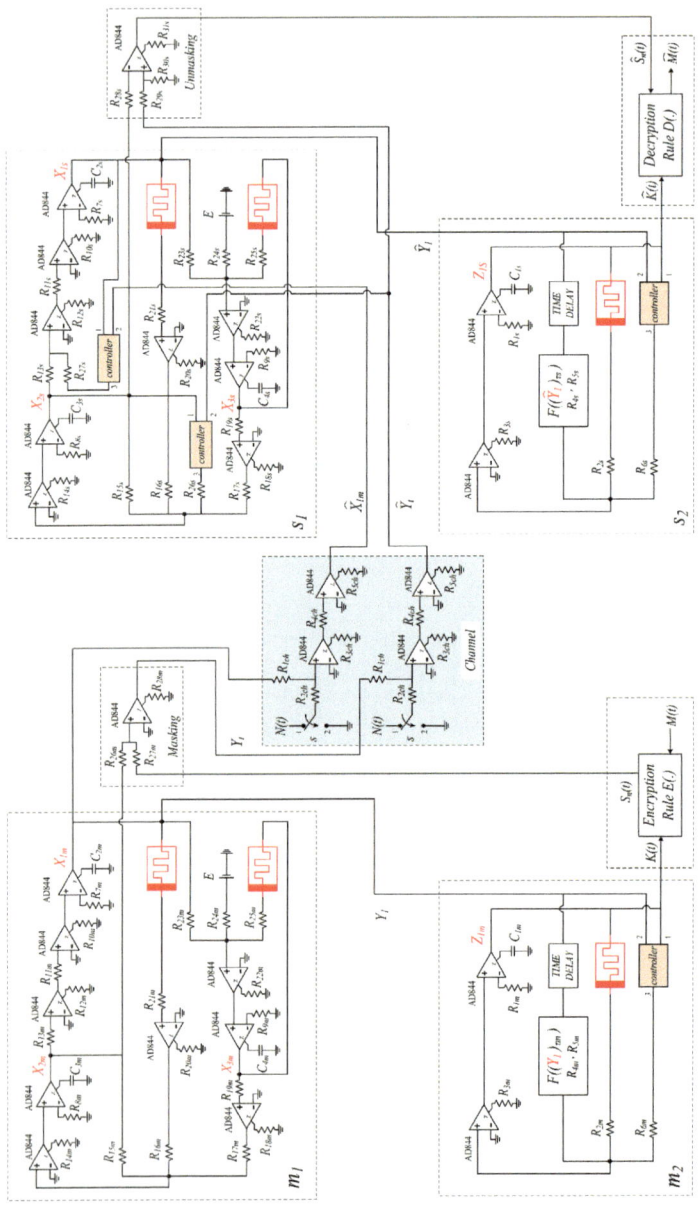

**Fig. 5.8** Circuit diagram for realizing the secure communication using memristive chaotic systems (5.4), (5.5), (5.6), and (5.7)

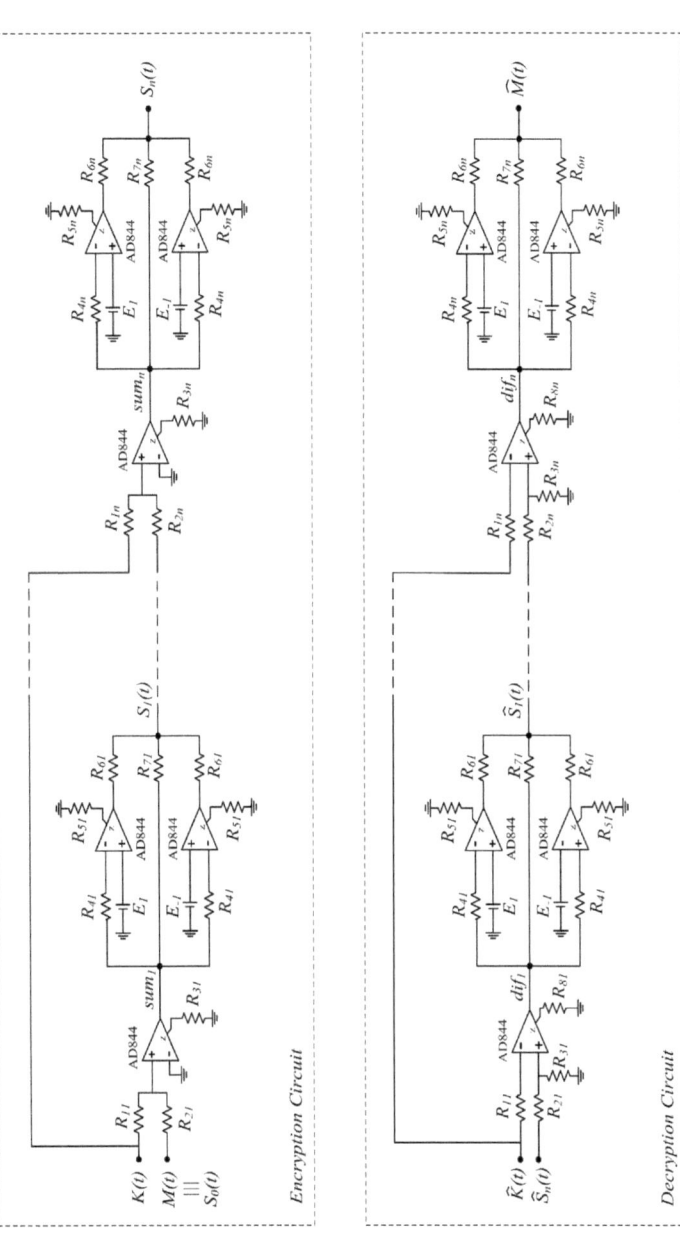

**Fig. 5.9**  Circuit diagram for realizing encryption and decryption rule of $n$-shift cipher. The components of circuits are selected as follows: $R_{1n} = R_{2n} = R_{3n} = R_{4n} = R_{7n} = R_{8n} = 1\,\mathrm{k}$, $R_{5n} = 100\,\mathrm{k}$, $R_{6n} = 10\,\mathrm{k}$, $n = 1, 2$. $E_1 = 1\,\mathrm{V}$, $E_{-1} = -1\,\mathrm{V}$. $V_{cc} = \pm 12\,\mathrm{V}$, $V_{\mathrm{sat}} = \pm 8.4\,\mathrm{V}$

**Fig. 5.10** Circuit diagram
for realizing controller parts,
and the components are
selected as: $R_{c1} = R_{c2} =$
$R_{c3} = R_{c4} = 1\,\text{k}$

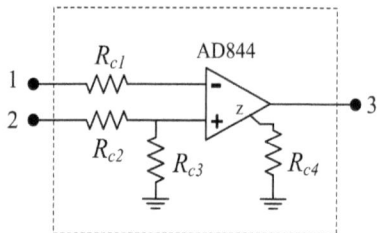

$$\hat{M}_n(t) = -\frac{|V_{\text{sat}}|}{R_{6n}}\,\text{sgn}\big[\text{dif}_n - E_j\big] + \frac{\text{dif}_n}{R_{7n}}, \quad j = \pm 1. \tag{5.11}$$

In masking and unmasking circuits, the $X_{2m}$ signal used for the masking process
and $X_{2s1}$ for unmasking as follows:

$$Y_t = -\left(\frac{R_{28m}}{R_{26m}}X_{2m} + \frac{R_{28m}}{R_{27m}}S_n(t)\right) \tag{5.12}$$

$$\hat{S}_n(t) = \left(\frac{R_{30s}}{R_{30s} + R_{29s}}\left(1 + \frac{R_{31s}}{R_{28s}}\right)\hat{Y}_t - \frac{R_{31s}}{R_{28s}}X_{2s}\right) \tag{5.13}$$

Figure 5.10 shows the circuit scheme of the controller blocks which illustrate in
Fig. 5.8. By applying Kirchhoff's laws to the designed electronic circuit, the nonlinear
differential equations are derived in the following form:

For the circuit $m_1$:

$$\dot{X}_{1m}(t) = \frac{1}{R_{7m}C_{2m}}\left(\frac{R_{10m}R_{12m}}{R_{11m}R_{13m}}X_{2m}\right)$$

$$\dot{X}_{2m}(t) = \frac{1}{R_{8m}C_{3m}}\left(\frac{R_{18m}R_{14m}}{R_{19m}R_{17m}}X_{3m} + \frac{R_{20m}R_{14m}}{R_{21m}R_{16m}}h(X_{1m}, X_{4m}) - \frac{R_{14m}}{R_{15m}}X_{2m}\right)$$

$$\dot{X}_{3m}(t) = \frac{1}{R_{9m}C_{4m}}\left(-\frac{R_{22m}}{R_{24m}}E - \frac{R_{22m}}{R_{25m}}h(X_{3m}, X_{5m}) - \frac{R_{22m}}{R_{23m}}X_{1m}\right) \tag{5.14}$$

For the circuit $m_2$:

$$\dot{Z}_{1m}(t) = \frac{1}{R_{1m}C_{1m}}\left(F\big((X_{1m})_{\tau_m}\big) - \frac{R_{3m}}{R_{2m}}h(Z_{1m}, Z_{2m})\right.$$

$$\left. + \frac{R_{3m}}{R_{6m}}(Z_{1m} - X_{1m})\right) \tag{5.15}$$

The state of switch $s$ is used for selecting the case of channel:

- Case 1: $s = 1$ (noisy channel), the received signal can be given as:

$$\hat{Y}_t = -\left(\frac{R_{3ch}}{R_{1ch}}Y_t + \frac{R_{3ch}}{R_{2ch}}N(t)\right) \qquad (5.16)$$

$$\hat{X}_{1m} = -\left(\frac{R_{3ch}}{R_{1ch}}X_{1m} + \frac{R_{3ch}}{R_{2ch}}N(t)\right) \qquad (5.17)$$

- Case 2: $s = 2$ (noiseless channel), the received signal equal to the transmitted one ($N(t) = 0$).

For the circuit $s_1$:

$$\dot{X}_{1s}(t) = \frac{1}{R_{7s}C_{2s}}\left(\frac{R_{10s}R_{12s}}{R_{11s}R_{13s}}X_{2s} + \frac{R_{12s}}{R_{27s}}\left(\hat{X}_{1m} - X_{1s}\right)\right)$$

$$\dot{X}_{2s}(t) = \frac{1}{R_{8s}C_{3s}}\left(\frac{R_{18s}R_{14s}}{R_{19s}R_{17s}}X_{3s} + \frac{R_{20s}R_{14s}}{R_{21s}R_{16s}}h(X_{1s}, X_{4s}) - \frac{R_{14s}}{R_{15s}}X_{2s} + \frac{R_{14s}}{R_{26s}}\left(\hat{Y}_t - X_{2s}\right)\right)$$

$$\dot{X}_{3s}(t) = \frac{1}{R_{9s}C_{4s}}\left(-\frac{R_{22s}}{R_{24s}}E - \frac{R_{22s}}{R_{25s}}h(X_{3s}, X_{5s}) - \frac{R_{22s}}{R_{23s}}X_{1s}\right) \qquad (5.18)$$

For the circuit $s_2$:

$$\dot{Z}_{1s} = \frac{1}{R_{1s}C_{1s}}\left(F\left((X_{1s})_{\tau_s}\right) - \frac{R_{3s}}{R_{2s}}h(Z_{1s}, Z_{2s}) + \frac{R_{3s}}{R_{6s}}(Z_{1s} - X_{1s})\right)$$

$$(5.19)$$

The components of circuits are selected as shown in Table 5.1.

Figure 5.11a, b shows the transmitted signal $Y_t$ and the error of keystream synchronization. It proves that the two memristive time-delay circuits $m_2$, $s_2$ are synchronized. Figure 5.12 illustrates the original message that had been transmitted through the noiseless channel with mismatch components and different initial conditions; also, the Figure shows the recovered message with and without filtering.

Figure 5.13a, b shows the transmitted signal $Y_t$ and the error of keystream synchronization. Figure 5.12 illustrates the original message that had been transmitted through the noisy channel (SNR $= 35$ dB) with mismatch components and different initial conditions; the Figure shows the recovered message with and without filtering Fig. 5.14.

Figure 5.15 shows the power spectra for the time waveforms of the transmitted signal $Y_t(t)$, keystream signal $k(t)$, and original message $m(t)$. Notice that, the original message $m(t)$ (yellow) is well-hidden in the main peak of the keystream signal $k(t)$ (blue), because the keystream is broadband (chaotic). The transmitted signal $Y_t(t)$ (red) has higher spike amplitudes and band from the keystream and original message signals, so this technique provided high security.

**Table 5.1**  Components of the circuit diagram

| Values | Components | |
|---|---|---|
| | $R_{im}$ | $R_{is}$ |
| 1 k$\Omega$ | $i = 11, 12, 13, 18, 19, 20, 21, 22, 23,$ 24, 25, 26, 27, 28 | $i = 11, 12, 13, 18, 19, 20, 21, 22, 23,$ 24, 25, 28, 29, 30, 31 |
| 10 k | $i = 1, 7, 8, 9$ | $i = 1, 7, 8, 9$ |
| 2 k | $i = 2, 3, 10, 14, 16$ | $i = 2, 3, 14, 16$ |
| 100 k | $i = 15$ | $i = 15$ |
| 20 k | $i = 17$ | $i = 17$ |
| 0.2 k | $i = 6$ | $i = 6$ |
| 0.02 k | | $i = 26, 27$ |
| 1.25 k | $i = 5$ | |
| 1.4 k | | $i = 5$ |
| 3 k | $i = 4$ | |
| 4 k | | $i = 4$ |
| 1.8 k | | $i = 10$ |
| | $C_{im}$ | $C_{is}$ |
| 100 nF | $i = 1, 2, 3, 4$ | $i = 1, 2, 3, 4$ |

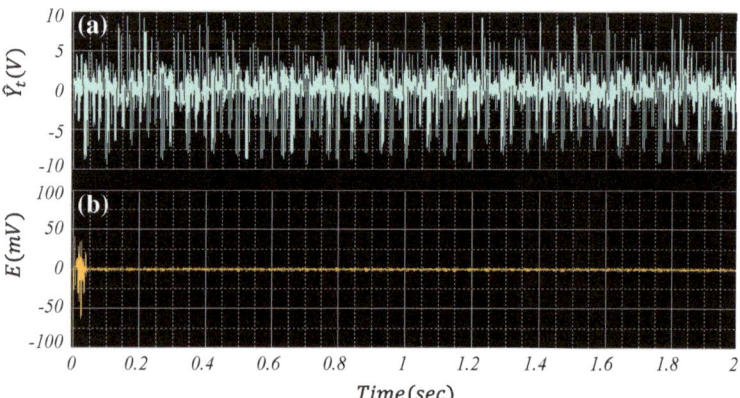

**Fig. 5.11**  PSpice results of chaotic secure communication mismatch components, feedback gains $C_0 = 100$, $l_2 = 10$, and noiseless channel **a** transmitted signal and **b** synchronization error in the estimation of the keystream

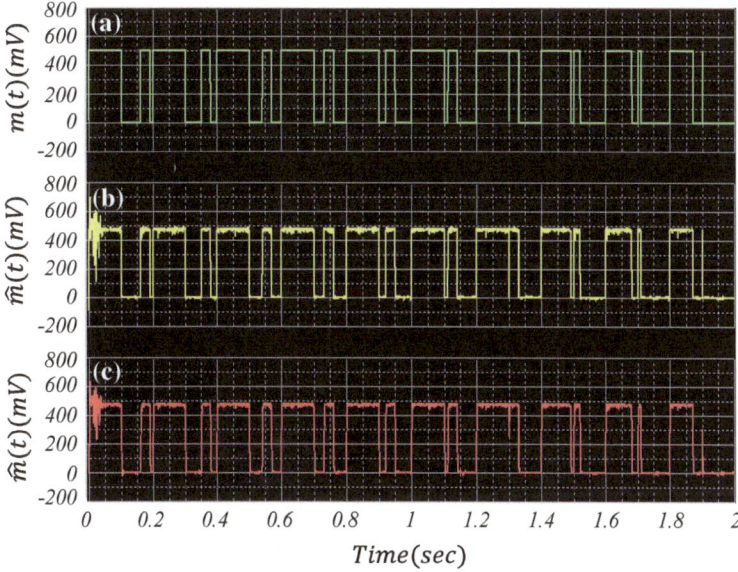

**Fig. 5.12** PSpice results of chaotic secure communication mismatch components, feedback gains $C_0 = 100$, $l_2 = 10$, and noiseless channel **a** transmitted message, **b** recovered message without filtering, and **c** recovered message with filtering

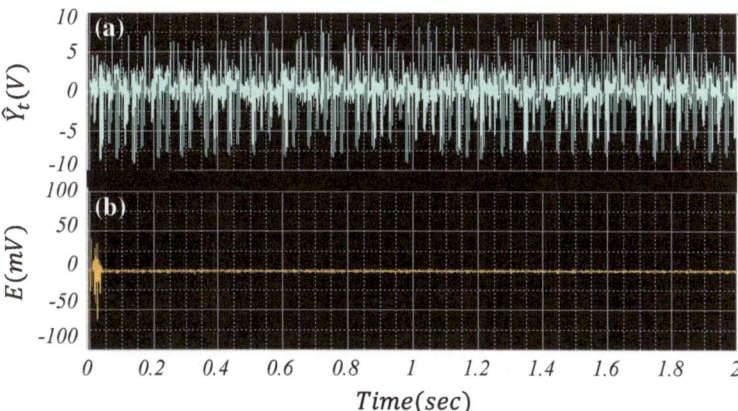

**Fig. 5.13** PSpice results of chaotic secure communication mismatch components, feedback gains $C_0 = 100$, $l_2 = 10$, and noisy channel **a** transmitted signal and **b** synchronization error in the estimation of the keystream

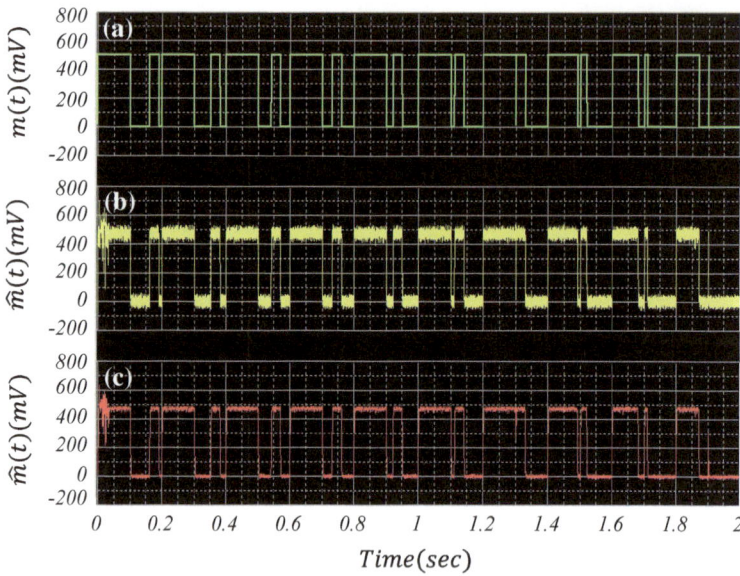

**Fig. 5.14** PSpice results of chaotic secure communication mismatch components, feedback gains $C_0 = 100$, $l_2 = 10$, and noisy channel (SNR $= 35$ dB) **a** transmitted message, **b** recovered message without filtering, and **c** recovered message with filtering

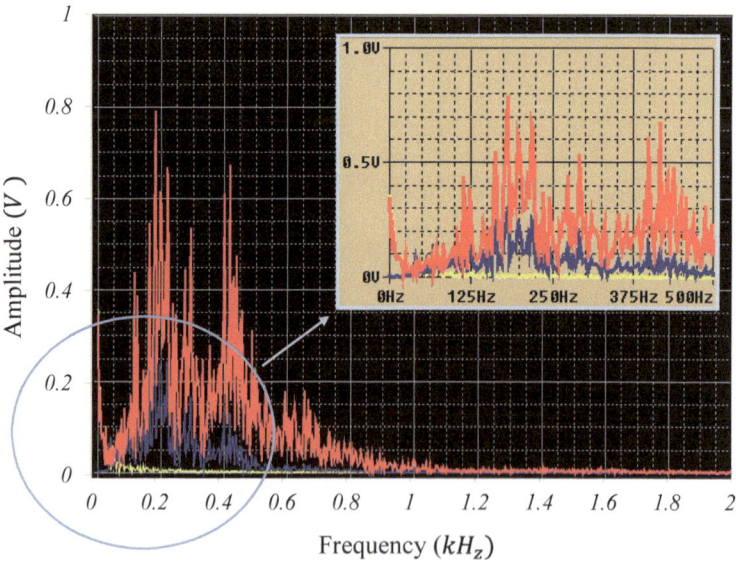

**Fig. 5.15** Power spectra of chaotic masking and message signals: transmitted signal $Y_t(t)$ (red), keystream signal $k(t)$ (blue), and original message $m(t)$ (yellow)

# References

1. K. Ganesan, R. Muthukumar, K. Murali, Look-up table based chaotic encryption of audio files. IEEE Trans. Circuits Syst. 1951–1954
2. T. Yang, C.W. Wu, L.O. Chua, Cryptography based on chaotic systems. IEEE Trans. Circuits Syst. I: Fundam Theory Appl. **44**(5), 469–472 (1997)

# Chapter 6
# Conclusions and Future Works

## 6.1 Conclusions and Contributions

The main goal of this book is to propose the memristor model and investigate its characteristics. The proposed memristor model requires less components for implementing its electronic circuit as compared to the proposed models in the research literature. The electronic circuit based on the proposed memristive systems is realized. The dynamical characteristics are investigated by phase portraits and bifurcation diagrams. The circuits have been implemented using current feedback operational amplifier (CFOA), so that they are suitable for high-frequency nonlinear oscillations. Good agreement between simulations results and the experimental observation has been found.

The main conclusions and contributions are summarized as follows:

1. New mathematical memristor model with a closed-form nonlinear function (inverse tangent) is proposed. The new function can be physically realized because the inverse tangent corresponds to the bipolar transistor differential pair.
2. Multi-scrolls attractors (complex behaviors) are generated from a memristive time-delay chaotic system using numerical simulation; then, the electronic circuit design is demonstrated using PSpice software. Later, proposed new memristive time-delay system, for generating two dimensions multiscroll attractor (grid). The new system that proposed by introducing another state to the time-delay chaotic system (3.2) is driven by staircase function.
3. New five-dimensional (5D) autonomous system with two memristors is proposed.
4. The effective synchronization is obtained using the PC and feedback control methods. In chapter four, the mentioned synchronization methods are described with the help of memristive time-delay dynamical circuit, which is considered as the circuits of the master (transmitter) and slave (receiver). In feedback control scheme, as feedback gain $C_0$ increases, the error of synchronization tends to zero. A noisy channel (AWGN) is used for checking the synchronization reliability with the noise effect where the added SNR range is 5, 10, 15, 20 dB.

© The Author(s), under exclusive license to Springer Nature Switzerland AG 2019
F. Rahma and S. Muneam, *Memristive Nonlinear Electronic Circuits*,
SpringerBriefs in Nonlinear Circuits, https://doi.org/10.1007/978-3-030-11921-8_6

5. The cryptography technique is proposed, which is used $n$-shift cipher algorithm for encrypt and decrypt message. Two types of message (analog: sound and digital: PN code) are introduced for secure communications. The results of the proposed cryptosystem are shown through the numerical simulations and the electronic circuit realization using PSpice software.

## 6.2  Future Works

1. Investigating an extent the notion of memristive model to inductive and capacitive element, which also displays *pinched hysteresis loop* in the voltage–current plane.
2. Investigating the relation between the memristor and the human brain. Leon Chua says "*Since our brains are made of memristors, the flood gate is now open for commercialization of computers that would compute like human brains, which is totally different from the von Neumann architecture underpinning all digital computers*". And Memristor-emulator synapses [1].
3. Investigating the design of memristor-based oscillators (second order: Wien oscillator and third order: phase shift oscillator) by demonstrating the possibility of sustained oscillation with pretenses of memristor (s).
4. The new time-delay chaotic system which displays grid attractor can be developed for image encryption.
5. Another type of the chaotic modulation could be treated as a basis to the wanted communication system.

## Reference

1. A.G. Radwan, M.E. Fouda, *On the mathematical modeling of memristor, memcapacitor, and meminductor*, vol 26 (Springer International Publishing Switzerland, 2015). ISBN 978-3-319-17490-7